Tommy Atkins

access to geography

ARID *and* SEMI-ARID ENVIRONMENTS

Michael Hill

D0293148

Hodder & Stoughton
A MEMBER OF THE HODDER HEADLINE GROUP

Acknowledgements

To all students and colleagues past and present who went on 'The Trip'.
All photographs by the author.

Every effort has been made to trace and acknowledge ownership of copyright. The publishers will be glad to make suitable arrangements with any copyright holders whom it has not been possible to contact.

Orders: please contact Bookpoint Ltd, 130 Milton Park, Abingdon, Oxon OX14 4SB. Telephone: (44) 01235 827720. Fax: (44) 01235 400454. Lines are open from 9.00–6.00, Monday to Saturday, with a 24-hour message answering service. You can also order through our website www.hodderheadline.co.uk.

British Library Cataloguing in Publication Data
A catalogue record for this title is available from the British Library

ISBN 0 340 800321

First Published 2002
Impression number   10 9 8 7 6 5 4 3 2 1
Year   2008 2007 2006 2005 2004 2003 2002

Cover photo: Scenery in Arches National Monument, Utah, USA (by the author)
Typeset by Fakenham Photosetting Ltd, Fakenham, Norfolk.
Printed in Great Britain for Hodder & Stoughton Educational, a division of Hodder Headline Plc, 338 Euston Road, London NW1 3BH by Bath Press, England.

Contents

Introduction

Approximately one-third of the Earth's land surface is classified as either arid or semi-arid, and this has had a fundamental impact on the global distribution of population and regional levels of economic development. The physical nature of arid lands is often perceived in a simplistic way; the portrayal of deserts in the media and advertising so often gives them the image of being hot and sandy places. However, within arid lands there are surprisingly great variations in all aspects of the physical environment; climate, ecosystems, soils, water availability and desert landscapes can all change within relatively short distances, emphasising the diverse nature of the arid environment.

The climates of arid and semi-arid regions are very distinctive and have a strong impact on plant and animal life, soil formation and the wide range of geomorphological processes at work within the desert landscape.

Desert hydrology is very different from that in more humid parts of the world. The balance between precipitation and evaporation gives deserts a water deficit throughout all or most of the year. Surface water sources are very limited in desert areas and have to be supplemented by various other sources, particularly groundwater. Inhabitants of arid areas rely heavily on irrigation for food supplies, and in the course of exploiting water resources encounter many problems, which may be environmental or political.

Plants and animals have a wide range of ways of adapting to aridity. Different deserts in different parts of the world have their own ranges of species, some unique to the particular area, others paralleled elsewhere. Desert soils are dry, often saline and poorly developed and therefore present numerous problems to those who need to use them for agriculture.

Weathering by both mechanical and chemical means, and the action of water and of wind are the main forces which mould the surfaces of arid regions. They work in different combinations, at different rates and over different time scales. It is this, together with the diversity of underlying lithology and geological structure which gives deserts their great variety of landscapes.

Desertification is currently a major world issue. There is much evidence which supports the idea of deserts increasing in area in some parts of the world. But, when studied over a much longer time scale, evidence suggests that deserts expand and contract as part of normal and perhaps cyclic patterns of climate change. In some places the growth of deserts appears to be reversible, in others it may not be so.

The economic potential of arid and semi-arid environments may be more limited than in the more temperate and humid parts of the world, but there are resources that allow desert regions to develop economically. For thousands of years nomadic pastoralism and oasis irrigation were the main forms of economy that sustained human populations in arid areas. In the modern world, a much wider range of economic activities takes place in deserts, notably those involving mineral production. The biggest economic transformations which took place in the twentieth century were those in the oil-rich countries, especially in the Middle East and North Africa.

1 The causes of aridity and the global distribution of deserts

'This huge desert is a mysterious and unfinished place, a trial run in landscape architecture where a few basic designs of sand, gravel and mountains are used repeatedly and stretched to their limits.

Swift (1975) *The Sahara*

1 Defining arid and semi-arid environments

The aridity of an area can be defined simply in terms of total annual rainfall, although climatologists have found it necessary to look at the relationship between rainfall and evaporation to give a more accurate definition, a matter which is discussed in Chapter 2. Basing definitions solely on rainfall totals, areas with less than 100 mm of rainfall per year are classified as being **hyper-arid**, whereas **arid** areas are generally taken to be those which receive less than 250 mm per year and **semi-arid** areas are those which have between 250 and 500 mm of rainfall per year.

2 The causes of aridity

Aridity is caused by a number of different factors working in combination; some of these are more significant than others, depending upon the individual desert location. The most important single factor which accounts for the global distribution of deserts is the **global pattern of atmospheric circulation**. The secondary factors contributing to aridity operate on a much more localised scale but are nevertheless of great significance, and these are the **rain shadow effect**, **continentality** and **cold ocean currents**.

a) The global pattern of atmosphere circulation

Most of the world's deserts lie in the tropical and sub-tropical belts between 20° and 30° North and South of the Equator. These latitudes are dominated by high pressure systems throughout most of the year. Figure 1.1 shows the mechanisms behind the build up of high pressure at these latitudes. Surface air converges close to the Equator at the **ITCZ (Inter-tropical Convergence Zone)** and rises up into the troposphere, it then moves towards higher latitudes and then mechanically sinks back down towards the Earth's surface at around latitudes 20°–30° North and South. These circulatory movements are known as the Hadley Cells. There are comparable circulatory movements resulting from ascending air in the temperate latitudes which moves towards the tropics and then descends, also around latitudes 20°–30°, forming the Ferrel Cells. As the air from both of these cells moves down towards the ground surface, high pressure is created by compression.

The high pressure belts are characterised by anticyclonic conditions for 90 per cent of the year. These bring clear skies, low rainfall and high rates of both insolation and evaporation. The almost continuous absence of cloud cover allows a great build up of heat during the day and considerable heat loss through radiation at night; it is therefore responsible for the high **diurnal temperature ranges**.

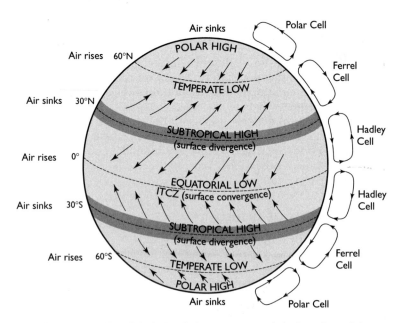

Figure 1.1 The global circulation pattern and the location of the subtropical high pressure belts

b) The rain shadow effect

Many arid and semi-arid parts of the world lie in the rain shadow of large mountain ranges. Where these mountain barriers are located close to the sea, they prevent moisture being carried by onshore air masses from reaching places located on their lee sides. Moist air which is being brought inland by prevailing winds or secondary winds will reach the mountain ranges and be forced to rise, leading to condensation and precipitation on the windward sides of the mountains. Warm, dry air then descends down the lee sides of the mountains which are consequently deprived of moisture.

Examples of the rain shadow effect in South America can be seen in

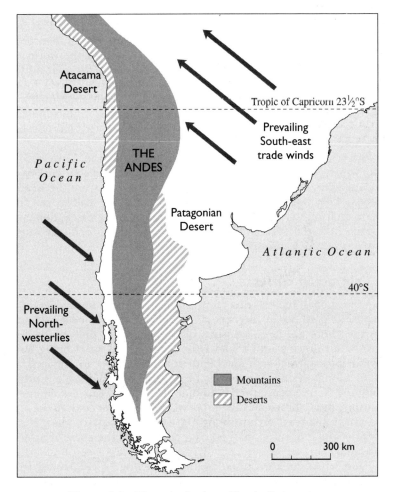

Figure 1.2 The rain shadow effect in S. America

Figure 1.2. In the tropical parts of the Southern Hemisphere, the main moisture-bearing winds are the prevailing south-easterly trade winds. The Great Dividing Range of Australia prevents moisture from being taken to the country's dry interior by these winds; the Andes create a similar barrier which prevents the south east trades from taking any moisture to the Atacama; and in South Africa the Drakensberg range stops moisture from reaching the Kalahari from the east.

In the Northern Hemisphere there are many examples of mountains acting as rainfall barriers: the Sierra Madre ranges in Mexico, the Sierra Nevada in the USA, the Aravalli hills in the Thar Desert of India and Pakistan, and the Elburz mountains in Iran.

c) The influence of continentality

Remoteness from the sea is another major cause of aridity. Places in maritime locations generally have a much higher rainfall than those in continental interiors. Some of the desert areas of the world are extremely remote from the sea, which helps to explain their aridity. The central parts of the Sahara are the most inland places to be found on the African continent, for example the Tassili Mountains of Southern Algeria are almost 2000 km from the sea. In locations the influence of continentality works in combination with latitude and rain shadow effect as a cause of aridity.

The regions most influenced by continentality are, however, the cold winter deserts of the Asian interior. Parts of the Turkestan, Taklamakan and Gobi Deserts are almost 2500 km from the nearest oceans. This not only helps to account for their aridity, but also means that given their high latitudes, and in some cases their high altitudes, these deserts suffer from very extreme temperature ranges between summer and winter.

d) Cold ocean currents

Some west coast located deserts in tropical latitudes have low rainfall because of cold ocean currents. As part of the Earth's heat exchange process, there are large transfers of energy within the oceans, with warm water moving into colder latitudes and cold water being transported into warmer latitudes. Four major cold ocean currents are found along western shorelines of deserts: the Benguela Current along the Namib coast, the Humboldt or Peruvian Current along the Atacama, the Californian Current off the South West coast of the USA and the Canaries Current along the Atlantic coast of the Sahara. There is also a less well marked cold current off the coast of western Australia.

These currents are caused by the upwelling of cold water from both great depths and from either the Arctic or Antarctic Ocean. The cold water influences the climate in two ways. First, moisture is con-

The hyper-arid coastline of the Atacama, near Antofagasta, N. Chile

densed offshore into fog and mist, which may then travel short distances inland only to be burnt off within a few hours of sunrise. Second, any onshore winds passing over the cold ocean surface will themselves be cool and therefore have a very low moisture carrying capacity, making them incapable of producing rain. Figure 1.3 illustrates how cold ocean currents influence coastal desert climates.

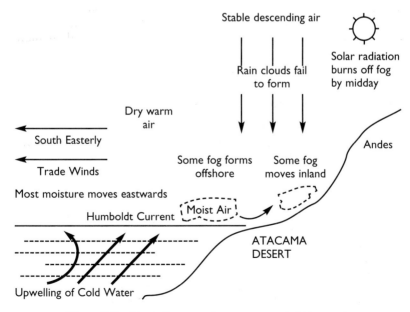

Figure 1.3 The influence of ocean currents (Atacama)

3 The global distribution of desert areas

There are five distinctive world desert zones, often referred to as provinces, which are clearly identifiable on Figure 1.4. The five zones are:

- The North African–Arabian–Central Asian province
- The Southern African province
- The North American province
- The South American province
- The Australian province

Each of these provinces has its own distinctive climates and ecosystems, as well as a different distribution of landscape types according to its geological structure and the processes which have moulded it through time.

4 The major deserts of the world

a) The North African–Arabian–Central Asian province

This province contains the largest area of arid land in the world, stretching from the Atlantic Ocean in the west, across the whole of North Africa, through the Arabian Peninsula into Iran and Pakistan and then branching up into Central Asia and China. It is a belt of some 15 000 km, encircling about one third of the Earth's circumference. Within this province are the Sahara, Arabian, Turkestan, Iranian, Thar, Taklamakan and Gobi Deserts.

i) The Sahara Desert

The Sahara Desert takes its name from the Arabic meaning 'empty area', and is the largest desert in the world. With an area of 9 million km², it covers over one-third of Africa and stretches 5000 km from the Atlantic Ocean in the west to the Red Sea and the Indian Ocean in the east. In the northern parts of the Sahara, country's names are often used to signify sub-divisions of the desert e.g. the Libyan Desert or the Algerian Desert. In the eastern part of the Sahara where it extends into the Horn of Africa, three regional names are given to the desert: the Danakil, the Chalbi and the Ogaden.

The Ethiopian Highlands create a rain shadow in the east and the Atlas Mountains prevent rainfall coming from the north. The cold current off the Canary Islands is responsible for the aridity along the Sahara's Atlantic coast, and the vastness of the desert ensures that the effects of continentality keep its interior dry. Above all, however, it is the location between latitudes 12° and 35° N, within the high pressure belt which gives the Sahara its year round aridity.

The countries which are completely or partially located within the Sahara are: Egypt, Libya, Tunisia, Algeria and Morocco (in North Africa), Chad, Niger, Mali, Mauritania (the countries of the Sahel or

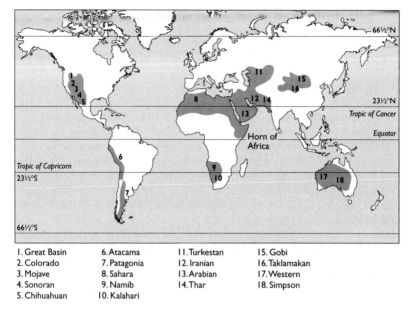

1. Great Basin	6. Atacama	11. Turkestan	15. Gobi
2. Colorado	7. Patagonia	12. Iranian	16. Taklamakan
3. Mojave	8. Sahara	13. Arabian	17. Western
4. Sonoran	9. Namib	14. Thar	18. Simpson
5. Chihuahuan	10. Kalahari		

Figure 1.4 The world distribution of deserts

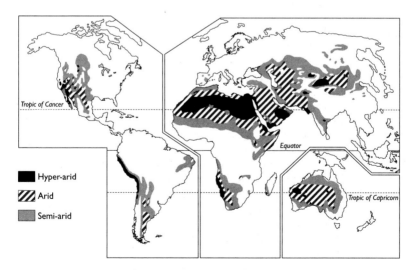

Figure 1.5 Arid, semi-arid and hyper-arid areas of the world

southern 'shore' of the Sahara), Ethiopia, Eritrea, Djibouti, Somalia, Kenya (in the Horn of Africa), the Sudan and the emergent state of Western Sahara. The Cape Verde Islands and most of the Canary Islands, are also parts of the Sahara.

ii) The Arabian Desert

The Arabian Desert covers a total of 2.6 million km², mainly contained within the triangular area of the Arabian Peninsula, but also stretching northwards as the Syrian Desert. The Arabian Desert is effectively an extension of the Sahara and exists for much the same reasons. Most of the desert is extremely dry, but the mountains of the south east along the Gulf have seasonal monsoon rains which annually top up the main aquifers of the peninsula. As well as Saudi Arabia, other countries within this desert include the UAE, Kuwait, Qatar, Bahrain and Iraq, all of which are oil-rich nations. The desert also encompasses Yemen and Oman on the southern tip of the peninsula and Jordan, Syria and Israel/Palestine to the north west.

iii) The Iranian Desert

The Iranian Desert lies almost exclusively within Iran and is a 'crossroads' between the hot deserts to the east and west and the cold winter deserts to the north. Covering an area of 300 000 km², this desert is mainly composed of a series of intermontane basins. Located between 24° and 31° N, it is within the high pressure belt and has extremely hot, dry summers. During these summer months there can be violent sandstorms during the so-called '120-day wind'.

iv) The Thar Desert

The Thar Desert is relatively small and covers an area of 175 000 km², straddling the southern part of the Indo-Pakistan frontier. The Aravalli Hills form the desert's south eastern edge, blocking out the monsoon and creating a rain shadow effect. Located between latitudes 23° and 31° N, the Thar lies totally within the high pressure belt.

v) The Turkestan Desert

The Turkestan Desert is one of Asia's three major cold winter deserts. Stretching from 34° to 51° N, most of it is beyond the influence of the high pressure belt. The main causes of aridity are the highly continental location and the rain shadow effects created by mountain ranges such as the Elburz in northern Iran. This desert is dominated by sand seas and the alluvial plains of dry rivers which drain to the Caspian Sea to the west and the shrinking Aral Sea. The three main countries which lie within this desert are Kazakstan, Uzbekistan and Turkmenistan, but it also extends into a small part of the Russian Federation.

vi) The Taklamakan Desert

The Taklamakan Desert is one of the most inhospitable places on Earth. The very name comes from the local Uigar word meaning 'labyrinth' or more accurately 'once you go in, you never come out'. The desert is dominated by huge shifting sand seas. Located between 35° and 42° N, the Taklamakan is hemmed in by the Tibetan Plateau to the south and the Tien Shan mountains to the north, and forms part of the great Tarim Depression which is 272 000 km^2 in area. Although summer temperatures are high, winter temperatures drop well below zero, due to the effects of continentality and high altitudes. The Taklamakan, which lies within the Xinjiang Province of China, is still widely unmapped.

vii) The Gobi Desert

The Gobi Desert has the most severe winter temperatures of any arid environment. It stretches from 35° to 49° N in the remote eastern interior of Asia, far away from any maritime influence and occupying a plateau which varies in altitude from 1000 to 3000 m. Both of these factors contribute to the extremely continental nature of its climate. Although the rainfall is generally less than 200 mm per year, during the summer months surface streams flow with water from the snowmelt in the neighbouring mountains. The Gobi lies within China and Mongolia.

b) The Southern African province

This region has two main desert areas, which, although adjacent to

The Gobi Desert – a cold winter desert – with part of the Great Wall of China, Jayuguan, W. China

each other are very different in character: the Namib and the Kalahari.

i) The Namib Desert

The Namib Desert stretches for 1200 km along the coastal belt of South West Africa between 15° and 35° S, extending 130 km inland. The name comes from the local dialect word meaning 'the land where there is nothing'. There are indeed vast areas devoid of vegetation, including 33 000 km² of sand seas such as the Sossusvlei sand dunes, which at 300 m are some of the highest in the world. The coastal plain of the Namib is hyper-arid, with just 30 mm of rainfall per year. A whole range of plant and animal life which is found there is adapted to collecting moisture from the sea fogs and mists created by the Benguela cold ocean current. These fogs have earned the coastline its nickname 'The Skeleton Coast' because of the large number of wrecked ships found along it. Further inland, rainfall reaches around 175 mm per annum in the mountains of the Great Escarpment. The vast majority of this desert lies within Namibia.

ii) The Kalahari Desert

The Kalahari Desert covers 570 000 km² of the interior of South West Africa, adjoining the Namib. The region is a mixture of high plateaux such as the Karoo and intermontane basins containing salt pans and large sand seas. The rainfall of the Kalahari varies from 460 mm in the semi-desert of the north, down to 130 mm where it merges with the Namib in the south west. Most of this desert lies within Botswana, but it also extends into Namibia and South Africa.

c) The North American province

The North American deserts are located in the USA and Mexico, and stretch for a wide latitudinal range from the Tropic of Cancer northwards to around 46° N in the Rocky Mountains. There are five distinct desert areas within this province.

i) The Great Basin Desert

The Great Basin Desert is almost 500 000 km² in extent and is the largest arid area in the USA as well as the most northerly (extending from 38° to 42° N). Geologically, it is formed from a series of parallel mountain ridges with deep valleys in between. The Great Salt Lake of Utah occupies the largest of these intermontane basins. The average annual rainfall varies throughout the region from 100 to 400 mm. The Great Basin Desert incorporates most of Nevada together with parts of California, Utah, Oregon, Idaho and Wyoming.

ii) The Mojave Desert

The Mojave Desert is the smallest desert in the USA, and extends for 140 000 km² between the latitudes 33° and 38° N. The Mojave is bounded by the Colorado River and the Sierra Mountains. Much of the region is high in altitude, especially in the west where the Paramint Ranges dominate the landscape. In the east these ranges give way to lower parallel ridges interspersed with dried up alkaline lake beds. The most dramatic of these is found in Death Valley, which at 150 m below sea level is the lowest point in the Western Hemisphere and experiences some of the highest temperatures on Earth in the summer months.

iii) The Colorado Desert
The Colorado Desert is comprised mainly of high plateaux and deep gorges, containing some of the most visited national parks in the USA. Stretching from 34° to 38° N, much of the area is composed of highly coloured sedimentary rocks, which accounts for the local name, the 'Painted Desert' (Colorado is the Spanish word for 'coloured').

Large areas are dominated by mesa, butte and canyon scenery, as found in the Grand Canyon and Mesa Verde National Parks and Monument Valley. Rainfall varies from 170 mm to 400 mm. The Colorado Desert extends into the states of Utah, Arizona and New Mexico as well as Colorado itself.

iv) The Sonoran Desert
The Sonoran Desert straddles the US–Mexican border and covers an area of 270 500 km². It borders the Pacific Ocean and extends down through the Baja California Peninsula. Much of the Sonoran is composed of internal drainage basins. The western part of the desert has a lower rainfall (around 50–100 mm), because of the rain shadow of the coastal ranges, whereas in the east rainfall may reach 250–300 mm as a result of seasonal thunderstorms originating in the Gulf of Mexico. One of the most distinctive features of the Sonoran are the huge stands of *saguaro* cactus.

v) The Chihuahuan Desert
The Chihuahuan Desert stretches over 450 000 km², mainly in Mexico, but with a small extension into South West USA and is located between latitudes 23° and 33° N. The two main mountain ranges, the Sierra Madre Oriental and Sierra Madre Occidental, create rain shadow effects from both the Atlantic and Pacific Oceans. Over 50 per cent of the Chihuahuan Desert is over 2000 m above sea level and it is mainly mountainous and rocky, but has salt pan accumulations within its internal drainage basins. One of the few areas of sand is the White Sands National Park in New Mexico, USA.

d) The South American province

South America has two great contrasting deserts. On the western side of the Andes is the Atacama, the closest desert to the Equator. On the eastern side of the mountain barrier is the cold winter desert of Patagonia.

i) The Atacama Desert

The Atacama Desert is almost 1000 km from north to south, yet is a mere 90–150 km wide. It stretches from 4° to 32° S along the coastal regions of Peru and Chile. The Andes create a rain shadow effect, preventing the easterly trade winds from reaching the Pacific coastlands. The Atacama is believed to be the driest place on Earth, and many places within it have had no rainfall since records began over 100 years ago. Several of the coastal towns only receive 10 mm of rain per year. For plant and animal survival rainfall is supplemented by moisture in the form of coastal fogs created by the presence of the Humboldt cold ocean current.

ii) The Patagonian Desert

The Patagonian Desert of Argentina also lies within the Andean rain shadow. As Patagonia is located at a much higher latitude than the Atacama, it lies within the prevailing westerlies wind belt and its position puts it on the leeward side of the Andes. Stretching from 37° to 51° S it is the most southerly desert in the world. Overall, its landscape is made up of a series of stone-scattered terraces which rise from the Atlantic coast up to the foothills of the Andes. Patagonia has a mean annual rainfall of between 100 and 200 mm, but as temperatures are relatively low for a desert, evaporation rates are not excessive and therefore scrub and steppe vegetation are abundant.

e) The Australian province

Australia's interior and western regions are largely desert; in fact Australia is the world's most arid continent. Referred to locally as the 'Outback', there are two distinctive subdivisions to Australia's central arid zone: the Western Desert and the Simpson Desert. Both of these areas extend between 18° and 32° S.

i) The Western Desert

The Western Desert covers an area of 650 000 km² and is made mainly of ancient plateaux averaging 300 m to 600 m above sea level. The desert has three separately named parts, the Great Sandy Desert, the Great Victorian Desert and the Gibson Desert, yet they merge into one another without being distinctly different. Several large mountain chains such as the Macdonell and Musgrave Ranges dominate parts of this desert and seas of sand dunes are common. In some areas ancient rock surfaces have been heavily eroded, leaving behind inselbergs and other types of relic mountains, which include Uluru or Ayer's Rock and the 36 domes known as the Olgas.

ii) The Simpson Desert

The Simpson Desert covers the eastern part of the Australian interior, and is generally lower lying than the Western Desert. There is considerable variety in its scenery, which includes ridges of parallel dunes 200 km long (the longest in the world) and a series of intermontane basins containing dried up salt lakes, of which the 19 500 km² Lake Eyre is by far the largest in Australia. Another characteristic of the Simpson Desert is the stone scattered 'gibber plains', which include the area known as the Sturt Stony Desert. Under the eastern part of the Simpson is the gigantic Great Artesian Basin, a 1.6 million km² store of fossil groundwater dating back to 20 000 years ago when the climate was much wetter.

Summary Diagram

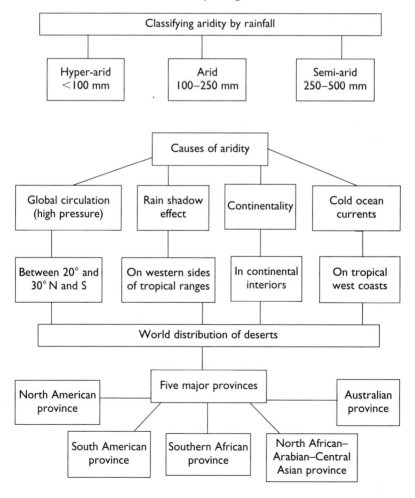

Questions

1. a Give a detailed explanation of the main causes of aridity experienced by the major desert regions of the world.

   b Taking several selected locations, explain why the causes of aridity can vary in importance from one part of the world to another.

2. a With the aid of sketch maps account for the distribution of the world's main desert areas.

   b Examine in detail why some desert areas are much more arid than others.

3. a With the aid of maps and diagrams explain how the following influence aridity:

   i cold ocean currents; **ii** the rain shadow effect; **iii** continentality.

   b Using the information on Figures 1.4 and 1.5, name three of the hyper-arid regions of the world.

   c Suggest reasons for the hyper-aridity of these regions.

4. a Select three desert areas from different continents and explain the main reasons for their aridity.

   b Using Figures 1.2 and 1.3, adapt these diagrams to explain the pattern of aridity in southern Africa.

   c Outline the main similarities and differences between the deserts of South America and southern Africa.

2 The climates of arid and semi-arid environments

'Fundamental to the creation of the characteristics of the arid environment is the climate. This conditions the way in which landforms, vegetation, animals, soils and modes of life differ in degree and kind from those in the humid areas of the earth's surface.'

Walton (1969) *The Arid Zones*

1 The characteristics and classification of arid climates

Aridity is the most important feature of a desert climate. As stated in Chapter 1, regions of the world located between the 500 and 250 mm isohyets are classified as semi-arid, regions between the 100 and 250 mm isohyets are regarded as arid and those experiencing under 100 mm of rainfall are regarded as hyper-arid. Aridity cannot, however be seen merely in terms of rainfall totals. Throughout the twentieth century, climatologists devised schemes for classifying climates by taking into account factors in addition to rainfall. Köppen brought out the first scientific classification of climate in 1918. Recognising the importance of plant growth in relation to what agriculture could take place, he built the concept of **precipitation effectiveness** into his system. Under Köppen's classification deserts are 'B' climates (the initial given to dry climates), which are sub-divided into 'BW' climates (arid) and 'BS' climates (semi-arid). Where mean temperatures are over 18 °C, the suffix 'h' is added and where they are lower than 18 °C,

the suffix 'k' is added. Thus low latitude hot deserts with a low rainfall
(e.g. the Sahara) are classified as 'BWh', whereas semi-arid high latitude cold winter deserts (e.g. Patagonia) are classified as 'BSk'.

In 1948, Thornthwaite modified the Köppen system by introducing
a P/E index. This involves dividing the total monthly precipitation
(P) by the total monthly evaporation (E), then adding the twelve
monthly values to get a grand total. In the Thornthwaite classification
the letter 'D' is used to designate semi-arid areas (those with a P/E
index of 16–31) and the letter 'E' is used for arid areas (those with a
P/E index of less than 16). He added lower case letters to indicate the
seasonality of the rainfall, which for arid and semi-arid areas were: 's'
for a summer deficiency, 'w' for a winter deficiency and 'd' for a year-
round deficiency.

In the 1970s Trewartha went back to the Köppen system and modified it; this was to become the classification system most commonly
used in atlases today. His group 'B' climates are those which have
evaporation exceeding precipitation and are summarised as follows:

- BWh: Hot deserts with annual mean temperatures over 18 °C.
        Related to tropical continental stable air masses and dry tropi-
        cal winds (e.g. the Sahara and Central Australia).
- BWn: Similar to the above but where there are frequent offshore fogs
        (e.g. the Namib and Atacama deserts).
- BWk: Mid-latitude interior deserts dominated by tropical continental
        air masses in summer, but by polar continental air masses in
        winter (e.g. the Turkestan and Patagonian deserts).
- BSh: Semi-arid tropical or sub-tropical areas. Tropical continental air
        masses are dominant, but there is a short rainy season (e.g. the
        Sahel region to the south of the Sahara and the Tell area of the
        Maghreb to the north of the Sahara).
- BSk: Semi-arid mid-latitude deserts with low summer rainfall (often
        from thunderstorms) and cold winters (e.g. the Gobi and the
        Great Basin deserts).

2 Temperature ranges and temperature extremes

Deserts have higher temperature ranges than more humid places at
comparable latitudes. These high ranges are both **diurnal** (between
day and night) and **annual** (between the hottest and coldest months).
During the day, incoming solar radiation is not impeded by cloud
cover and temperatures are allowed to rise rapidly; the reverse hap-
pens at night when the clear skies allow long-wave radiation to escape
from the ground surface and temperatures fall quickly. Diurnal ranges
of 15–20 °C are common in deserts, both in summer and winter
months. Places closer to the sea will generally have a lower diurnal
range than this, whereas locations deep in the continental interiors will

have much higher ranges. Also, the more elevated the desert location, the greater the difference in temperature between day and night. Latitude is the single most important factor which determines the differences between summer and winter maxima. Tropical deserts can have ranges of less than 10 °C, whereas in cold winter deserts they may be in excess of 30 °C.

The highest shade temperatures normally recorded in deserts are around 45–55 °C, although 58 °C has been recorded in the Libyan Desert and 57 °C in Death Valley in California. Actual surface temperatures (not in the shade) are much higher, the most extreme example having been recorded in the Red Sea Hills of Egypt at 82.5 °C. This is because 'in the shade' temperatures record the temperature of the air rather than of rock or sand surfaces. **Albedo** or the reflectivity of a surface colour, has a big influence on surface temperatures. Salt lakes and salt crusts which are white or very light in colour reflect 40–80 per cent of incoming radiation and therefore have surface temperatures which can be as much as 10–15 °C lower than areas with dark surfaces, such as landforms made of basalt, which reflect only 5 per cent of incoming radiation.

3 Rainfall totals and rainfall variability

Desert rainfall is both low and variable to the degree of being unreliable. Semi-arid locations tend to have greater seasonality in their rainfall patterns than either arid or hyper-arid regions, because they partially reflect the climates of the regions which they adjoin.

As a general rule, the lower the annual rainfall, the less reliable it is. Annual statistics are averaged out over at least a 30 year period in order to be significantly accurate. This works well in humid parts of the world, but in hyper-arid regions a 30 year total may be the product of just a few rainfall events.

The reliability of rainfall is calculated and expressed in the **rainfall variability index**, the formula for which is as follows:

$$\text{variability (per cent)} = \frac{\text{the mean deviation from the average}}{\text{the average}} \times 100$$

Rainfall variability is only around 5 per cent in very wet parts of the world such as monsoon Asia; around 10 per cent in cool temperate countries such as the UK; and around 15 per cent in the countries of Mediterranean Europe. In semi-arid and arid regions the index goes much higher; for example in Karachi, Pakistan it reaches 30 per cent and in Alice Springs in Australia it reaches 60 per cent. The most extreme example on record is the Dakhla Oasis in Egypt, which has an index of 150 per cent; this town has an average annual rainfall of 0.5 mm which makes it difficult to calculate the index in the first place. In parts of the Atacama Desert in Chile, such as the town of Arica, rainfall hardly ever occurs and it is impossible to calculate the variability index; if it could be calculated it would be well in excess of 200 per cent. Figure 2.1 shows rainfall variability in Namibia.

4 Extreme rainfall events

Although most desert storms are of low intensity, there are occasional spectacular storms which reflect the high variability index found in arid regions. Some of the extreme rainfall events during the twentieth century are shown opposite.

As can be seen from Figure 2.2, it is possible for a desert location to receive more than its average annual rainfall in just one day. In the Sahara, violent thunderstorms are the main cause of extreme rainfall events. In the case of Lima, however, the immensely heavy storm was due to an **El Niño** event. Approximately every ten years, around late December and early January, the normal circulatory pattern of currents in the Pacific Ocean changes. The cold Humboldt Current which is responsible for keeping the coastlines of Peru and northern

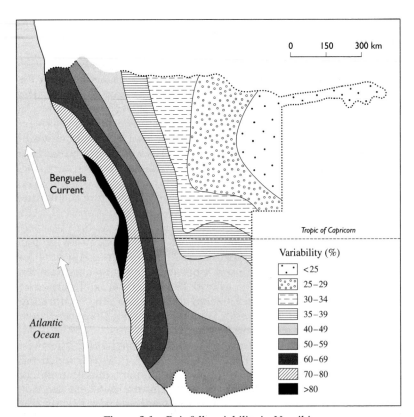

Figure 2.1 Rainfall variability in Namibia

Chile arid, is replaced by a warm current which brings with it very heavy rainfall. This current is known as 'El Niño' or 'little baby', a reference to the fact that it comes at Christmas time.

The consequences of extreme rainfall events such as these are flooding, which may cause damage to towns, villages, infrastructure, crops and livestock. As a large proportion of the storm water is likely to be in the form of rapid sheet runoff, it will not be trapped and stored for human use.

Location	Average annual rainfall (mm)	Extreme storm event (mm)
Lima, Peru	27	1524 in 24 hours
El Djem, Tunisia	275	319 in 3 days
Bisra, Algeria	148	210 in 2 days
Djibouti, Djibouti	129	211 in 24 hours
Swakopmund, Namibia	15	50 in 24 hours

Figure 2.2 Extreme desert rainfall events of the twentieth century

5 Evapotranspiration

Evapotranspiration rates are very high in arid regions because of the lack of cloud cover and high rates of incoming solar radiation. The low density of plant cover in arid areas means that the vast majority of water loss involved in evapotranspiration is from evaporation rather than transpiration. **Potential evapotranspiration (PE)** is the amount of evapotranspiration that would take place if water sources are unlimited. Rainfall is low in arid and semi-arid environments and needs to be supplemented by irrigation water to sustain plant growth. At the same time surface stores of irrigation water e.g. in reservoirs and canals can be depleted quickly by high rates of evaporation. The measurement of potential evapotranspiration rates is therefore very important in deserts within the context of agricultural potential and the development of new farming areas. Various devices are used for this measurement, including the **evaporation pan** and the **lysimeter**. Evaporation pans are shallow, cylindrical metal pans containing water from which the rates of water loss can be monitored. Lysimeters are cylinders set in the ground, within which the rates of water loss from a vegetated column of soil are measured.

The huge spatial differences between rainfall and potential evapotranspiration can be seen in Figure 2.3, which shows the two phenomena for continental Australia. The highest potential evapotranspiration rates are in the central part of Australia, where the rainfall is lowest. It is also noticeable that in some parts of the Australian Desert, PE rates are as much as 30 times as great as the local rainfall.

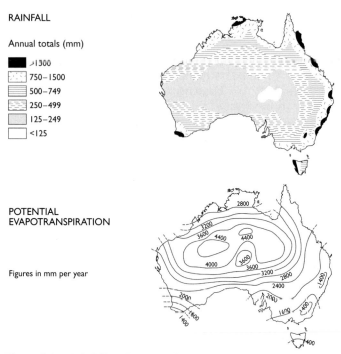

Figure 2.3 Rainfall and potential evapotranspiration in Australia

6 Winds in arid environments

Winds are another important aspect of desert climates. Approximately 20 per cent of desert surfaces are composed of loose sand or smaller particles, so wind has a major role in the moulding of these landscapes.

Most deserts are influenced by strong local and seasonal winds. Within and around the Sahara there are numerous local winds, some are hot, whereas others are cooler and sometimes humid. The **harmattan** blows down through the Sahel into Nigeria and other countries along the Gulf of Guinea during the November–April dry season; this hot, dry wind is often accompanied by dense dust storms. The **khamsin** blows northwards down through the Nile Valley in Egypt and brings high temperatures and dust storms to Cairo and the Nile Delta, especially during the spring. The **sirocco** and the **leveche** are both winds which have their origins in North Africa but bring warm and often muggy conditions and dust storms to Italy, Spain and the South of France. Within the western Sahara, under normal conditions, the **alisio** wind blowing from the north creates relatively cool conditions, however in the spring the **irifi** wind may blow from the east bringing

intense heat and dust from the Sahara. In Tunisia there are three
types of local depression winds: the **chelili** which blows from the south
in spring and summer, bringing hot, dry conditions from the Sahara,
the **djebeli**, which occasionally brings cool descending air from the
Atlas mountains and the **chegui** which blows from the Mediterranean,
bringing humid air to the coastal areas, especially in the winter
months.

7 The climates of selected weather stations

Figure 2.4 shows the climatic statistics for 24 selected meteorological
stations located in deserts throughout the world. The stations are
listed according to their latitudes, moving away from the Equator.
What becomes clear as the statistics are analysed, is that there is no
such thing as a 'typical desert climate', as each location is unique in
its local deviations from the norm. In addition to latitudinal location,
altitude, distance from the sea, coastal currents, winds and the proximity of mountain ranges all account for the variations which can be
found even over relatively short distances.

Lobitos and Lima both have the very low rainfall associated with
the Atacama Desert. Lobitos is at an Equatorial latitude, which would
not normally be so arid, but the Humboldt Current influences the climate in this locality. The Equatorial latitude is responsible for the very
low temperature range. Lima's temperatures are low for such a tropical latitude because of the prevalence of coastal fogs for 80 per cent
of the year.

Aden, on the southern tip of the Arabian Peninsula, has a similarly
low rainfall to the Atacama locations and a similar latitude to that of
Lima, but in the absence of a cold ocean current and the consequent
coastal fogs, the temperatures are much higher than in the Peruvian
stations, especially in the summer months.

Timbuktu in Mali is strictly speaking a semi-desert location
(despite its 210 mm rainfall total) because of its position on the edge
of the dry savanna region to the south of the Sahara; it therefore has
a marked rainy season in its summer months, which also explains why
its hottest period is in the late spring, before the rains arrive. Hall's
Creek in Australia is very much a Southern Hemisphere equivalent of
Timbuktu, as it is in a similar type of location. The slightly higher latitude and the higher elevation are responsible for its lower temperatures and higher rainfall.

Jiddah in Saudi Arabia has prolonged drought for much of the
year and its rainfall is concentrated in the three winter months when
depressions are most likely to bring rain in from the Red Sea or
beyond. The high summer temperatures reflect the period when the
weather is dominated by hot winds blowing from the interior of the
Arabian Peninsula. Sanluis Potosi in Mexico has a high elevation, that
gives it a semi-desert level of rainfall, which comes mainly in the form

of summer thunderstorms. The higher altitude also moderates temperatures, making them lower than they would normally be at that latitude.

Walvis Bay in Namibia has low temperatures for its latitude as a result of the coastal fogs and on-shore winds associated with the Benguela Current. The low rainfall is typical of a hyper-arid region dominated by a cold ocean current. Alice Springs, close to the centre of the Australian Outback has an annual rainfall total of 246 mm, which puts it on the margin between desert and semi-desert. Its high elevation and the prevalence of summer thunderstorms account for the slightly higher rainfall than would normally be experienced at this latitude. Antofagasta has conditions which are similar to those at Walvis Bay. With its west coast Atacama location and the influence of the Humboldt Current, it experiences offshore fogs, has low temperatures for a desert city and a hyper-arid level of rainfall. By contrast, the position of Aswan in the interior of Egypt, puts it a long way from any maritime influences and therefore it has an exceptionally low rainfall. Any potential moisture carried by winds from the west is lost over the Red Sea Highlands.

Karachi, on the edge of the Thar Desert in Pakistan has its maximum rainfall at the same time as the Indian monsoons, when some rain manages to get across the mountain barriers to the east; this is also the hottest period when isolated thunderstorms may be generated with moisture from the Arabian Sea.

In Salah and Cape Juby are both Saharan locations at much the same latitude, yet they have different climates. In Salah has a continental location, close to the centre of the Sahara and therefore has a more extreme climate; its summers are very hot and the winters are mild, giving it a high temperature range. Its winter rainfall is negligible. By contrast Cape Juby is on the coast of western Sahara, where its temperature range is greatly modified by the influence of the offshore cold Canaries Current. The local 'alisio' winds blowing down from the Atlas mountains also help to moderate the climate.

Yuma, Arizona and San Diego, California are at similar latitudes but their temperature ranges are very different due to their positions in relation to the sea. Yuma is very distant from the sea and as a consequence has hot summers and cool winters. San Diego is on the coast where it is influenced by the cool offshore California Current. It therefore has a much narrower temperature range. The higher rainfall total is due to its being at a latitude where winter depressions come in from the Pacific.

Tehran is located on a high plateau close to the Elburz Mountains. Given its latitude, this 'physical' location is responsible for both cold winters and a distinctive winter maximum to its rainfall. Mosul in northern Iraq is at a similar latitude but a lower altitude; it therefore has a more prolonged summer drought. Its higher winter rainfall is due to the fact that it is occasionally under the influence of depression tracks at that time of the year.

Death Valley, California has a similar climate to neighbouring Yuma, Arizona. Its location in a deep intermontane basin below sea level gives it a strong rain shadow effect, although there can be summer thunderstorms, which give it a summer maximum. The 'frying pan' effect of this basin is also responsible for some very extreme summer temperatures for its latitude. Ashkhabad, Kashgar and Deaver, Wyoming are all high latitude desert locations. All three of them have huge temperature ranges with sub-zero conditions in the winter months, this being the result of their locations within continental interiors. Their rainfall patterns are widely divergent due to different local topography and wind patterns. Ashkhabad has a well-marked summer drought, Deaver has a spring and summer maximum, whereas Kashgar has a much more erratic rainfall pattern.

Commodoro Rivadavia in Patagonia has a very low temperature range for its latitude, due to its proximity to the sea and the influence of strong and constant prevailing winds. Rainfall is evenly distributed throughout the year, also as a result of the maritime influences and strong winds. Astrakhan, in the Russian Federation has very similar climatic conditions to the other cold winter interior desert stations. It has severe winters and hot summers because of its continental position.

Summary Diagram

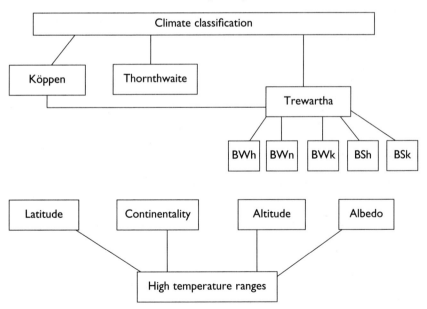

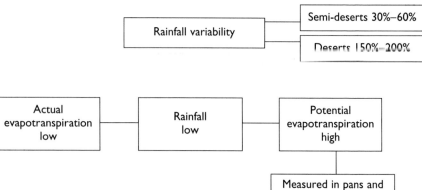

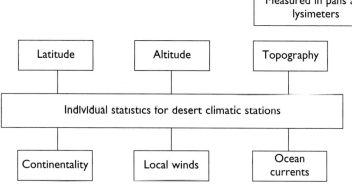

Questions

1. a What are the main characteristics which make desert climates so distinctively different from those in other parts of the world?

b Using specific examples of desert locations, examine the contention that 'there is no such thing as a typical desert climate'.

2. a Examine the major differences between the climates of hot deserts and cold winter deserts; and between semi-arid and hyper-arid locations.

b Why is rainfall so variable from one year to the next in deserts and why is it so variable between different desert locations in the world?

3. a Explain what is meant by the rainfall variability index and why it is useful.

b Use Figure 2.4 to compare and contrast the annual rainfall patterns of:

i Timbuktu and Walvis Bay.

ii Yuma and Ashkhabad.

c Using Figure 2.4 examine what relationships may exist between altitude and rainfall totals.

4. **a** With reference to Figure 2.4, and supporting your answer with statistics, explain how latitude influences the temperature range in desert locations.

 b Choose three contrasting climatic stations from Figure 2.4 and sketch out climatic graphs for each. Annotate these graphs with their most important characteristics.

 c Outline the main reasons for the contrasts between their climates.

STATION	LATITUDE	ELEVATION (M)		J	F	M	A	M	J	J	A	S	O	N	D	TEMP. RANGE	RAINFALL TOTAL
Lobitos, Peru	4°27'S	22	T°	26	27	27	26	24	22	22	21	20	22	22	23	7	50
			R mm	10	27	3	0	0	0	0	0	0	0	0	0		
Lima, Peru	12°02'S	115	T°	22	22	22	20	18	16	16	16	16	16	17	19	7	27
			R mm	0	0	0	0	3	3	3	5	5	3	0	0		
Aden, Yemen	12°45'N	10	T°	24	25	26	28	31	32	34	34	31	29	27	25	8	52
			R mm	8	5	13	5	3	3	3	3	5	3	3	3		
Timbuktu, Mali	16°46'N	260	T°	22	25	28	32	34	34	32	30	31	32	28	23	12	210
			R mm	0	0	0	0	3	20	34	81	24	3	0	0		
Hall's Creek, Australia	18°13'S	366	T°	30	29	28	26	21	19	19	21	24	28	31	31	13	450
			R mm	137	107	71	13	5	5	5	13	2	13	35	79		
Jiddah, Saudi Arabia	21°03'N	6	T°	23	25	27	32	34	34	34	30	31	32	28	23	11	77
			R mm	22	0	0	0	0	0	0	0	0	0	0	15		
Sanluis Potosi, Mexico	22°09'N	2000	T°	13	15	17	21	22	21	21	19	18	17	15	14	9	349
			R mm	12	19	8	5	30	72	34	55	86	17	10	7		
Walvis Bay, Namibia	22°50'S	8	T°	19	19	19	18	17	16	16	14	14	15	17	18	4	25
			R mm	3	5	9	3	3	0	0	3	4	0	0	0		
Alice Springs, Australia	23°38'S	600	T°	29	28	25	20	16	12	12	14	18	23	26	28	17	246
			R mm	42	33	28	10	15	12	8	8	8	17	30	38		
Antofagasta, Chile	23°29'S	100	T°	21	21	19	17	16	14	14	14	14	16	17	19	8	9
			R mm	0	0	0	0	0	0	0	3	0	0	0	0		
Aswan, Egypt	24°02'N	130	T°	16	17	21	26	31	33	33	33	31	28	23	17	15	3
			R mm	0	0	0	0	0	0	0	0	0	0	0	0		
Karachi, Pakistan	24°48'N	4	T°	19	20	24	27	30	31	30	28	28	27	23	20	22	194
			R mm	12	10	8	3	3	17	81	40	12	0	3	5		
In Salah, Algeria	27°12'N	280	T°	13	16	20	25	29	35	37	36	33	27	19	14	24	15
			R mm	2	3	0	0	3	0	0	0	0	0	3	5		
Cape Juby, W. Sahara	27°56'N	6	T°	17	17	17	18	19	20	21	21	21	21	19	17	4	49
			R mm	8	5	8	0	0	0	0	0	8	0	3	8		
Yuma, Arizona USA	32°40'N	65	T°	13	16	19	23	27	31	35	34	31	24	18	14	22	85
			R mm	8	8	5	3	0	0	0	8	8	8	3	15		
San Diego, Calif. USA	32°43'N	6	T°	12	13	14	16	17	19	21	21	21	19	15	13	8	246
			R mm	46	48	38	15	5	3	0	0	8	17	23	46		
Tehran, Iran	35°44'N	1219	T°	1	6	9	16	22	27	31	30	25	19	11	6	28	240
			R mm	46	25	46	36	13	3	3	0	3	8	20	33		
Mosul, Iraq	36°19'N	222	T°	7	9	13	19	23	29	29	28	25	21	15	9	25	384
			R mm	71	78	54	36	18	0	0	0	3	9	48	61		
Death Valley, Calif. USA	36°38'N	-54	T°	11	13	19	24	29	35	37	36	32	24	19	14	26	40
			R mm	2	13	2	3	5	0	3	8	0	3	3	3		
Ashkhabad, Turkmenistan	37°57'N	226	T°	-1	0	8	15	19	26	29	24	22	17	8	4	30	226
			R mm	25	20	8	35	30	5	3	3	3	13	20	18		
Kashgar, China	39°30'N	1297	T°	-6	0	8	16	21	25	27	24	21	13	4	-3	33	97
			R mm	8	7	8	5	20	10	3	3	3	13	0	5		
Deaver, Wyoming, USA	44°53'N	1300	T°	-8	-5	0	5	13	20	23	18	14	8	-1	-6	30	133
			R mm	5	3	10	17	23	17	23	21	8	10	5	5		
Commodoro Rivadavia, Argentina	45°47'S	65	T°	19	18	16	13	9	8	8	9	10	13	16	18	12	205
			R mm	10	15	17	17	30	25	25	18	14	9	2	12		
Astrakhan, Russia	46°15'N	-15	T°	-7	-5	-1	9	18	23	25	23	17	9	2	-3	32	161
			R mm	12	12	10	15	15	20	12	10	15	10	15	15		

3 The desert hydrological cycle and water problems

KEY WORDS

Water balance or budget: a graph showing the annual pattern of rainfall and potential evapotranspiration.

Perennial river: a river that flows throughout the year.

Exogenous river: a river flowing into a desert which has its origin outside.

Wadi: a dry desert watercourse which only flows with water following rainfall.

Aquifer: a water bearing rock.

Desalination: the process whereby salt is removed from sea water.

Endoreic or internal drainage: drainage which flows into an inland basin rather than to the sea.

'Clearly the most precious resource of the arid lands is water. For vegetation, animals and man it controls their very existence, their distribution and their density.'

Walton (1969) *The Arid Zones*

1 The hydrological cycle in arid environments

Figure 3.1 is an idealised version of how the hydrological cycle operates in arid environments. Rainfall is low and sporadic. Water from light storms percolates down to aquifers, although some is lost in evaporation. During the more intense and heavy storms, there is much greater overland flow and therefore higher evaporation. **Actual evapotranspiration (AE)** rates are almost equivalent to the amount of rainfall which stays on the surface where impermeable soils and rocks prevent infiltration from taking place. By contrast **potential evapotranspiration (PE)** is much higher, often as much as ten times that of the annual precipitation total. The highest potential evapotranspiration rates are found within basins which act as heat traps. Figure 3.2 explains the vast differences which exist between actual and potential evapotranspiration within an arid environment.

Water from aquifers is used for irrigation in oases, reaching the surface either naturally through springs or artificially through wells. Annual variations in rainfall totals greatly influence the mechanisms

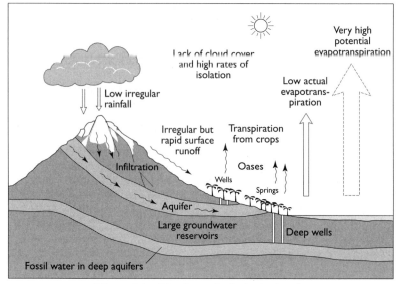

Figure 3.1 The desert hydrological cycle

within the desert hydrological cycle: they affect rates of evaporation and levels of infiltration, which then cause fluctuations in the water table, which in turn have an impact upon oasis cultivation and food supplies.

2 The water balance

Water balance or **Water budget** graphs are a very clear way of examining the relationships which exist within different climatic regions between their annual rainfall and evapotranspiration patterns and how these influence the need for irrigation. The annual water balance for arid and semi-arid regions follows a distinctive pattern. The limited rainfall is, on balance, greatly outweighed by evapotranspiration levels. Most desert locations find themselves with a **water deficit** for most, if not all, of the year. This poses great problems of water supply for agriculture, domestic and industrial use. In most arid countries, irrespective of whether they are LEDCs or MEDCs, it is generally agriculture which makes the greatest demands on local water supplies. However, growing populations and the process of urbanisation are causing an ever increasing strain on water supplies in cities. The development of irrigation schemes in the countryside and aqueduct supply routes to cities have a long history in predominantly arid areas such as the Middle East, North Africa and Andean South America. In these regions, hydraulic engineering schemes have enabled great civilisations to flourish.

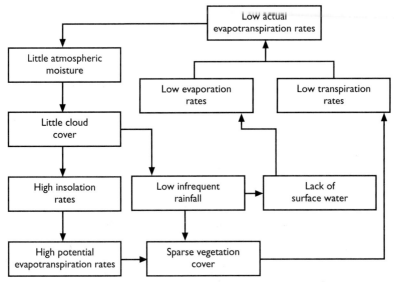

Figure 3.2 Differences between actual and potential evapotranspiration in arid areas

The four water balance graphs (Figure 3.3) are for urban areas in the Middle East and North Africa. Although the graphs are similar in overall form, there are big differences between the climates the four cities experience. Beirut in the Lebanon has a dry Mediterranean climate, with an annual rainfall total of 517 mm. Its limestone aquifers are recharged throughout the wetter winter months, when evapotranspiration rates are lower. The soil moisture is only used up in May and the summer drought begins in June, marking the beginning of when irrigation is needed by farms in the surrounding area. Tripoli in Libya also has a coastal Mediterranean location, but with the lower rainfall of a semi-arid area (384 mm per annum) and higher evaporation rates, it uses up its surplus winter rainwater supplies by April. Tripoli thus has a longer period of summer drought, requiring a greater use of irrigation to make up for the moisture deficit. Baghdad, with its inland location has an annual rainfall total of just 125 mm, making it much drier than Tripoli; its summer temperatures are also much higher, creating a greater moisture deficit. Irrigation water is needed for most weeks of the year in the farms and smallholdings around Baghdad. The Cairo region has no chance of replenishing its local aquifers from the meagre and unreliable winter rainfall which averages out to a mere 24 mm per annum. Without the abundant water supplies from the Nile, Cairo and its rich agricultural hinterland would not exist.

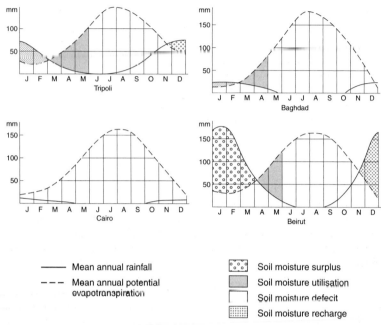

Figure 3.3   The water balance for four arid and semi-arid locations

3 The sources of water in arid environments

Although water supplies are very limited in arid and semi-arid areas, the precious water needed to support human life is taken from a variety of different sources. Which of these is used in which desert depends upon the local hydrological resources, geological structure, proximity to the sea and levels of national wealth and technological knowhow. The eight main sources of water supply in desert areas are as follows.

a) Large perennial rivers

Large perennial rivers are extremely important in giving life to what would otherwise be very barren areas. They are generally **exogenous rivers**, i.e. those which have their origins outside of an arid zone but flow through it. The River Nile in Egypt which has its origins in the Ethiopian Highlands; the Rivers Tigris and Euphrates in Iraq which have their sources within the Anatolian Plateau of Turkey; and the Colorado River in South West USA which rises in the higher parts of the Rocky Mountains are all examples of major exogenous perennial rivers. All of these rivers are highly managed with dams, reservoirs and irrigation canals, reflecting their great importance, nationally and internationally.

b) Small perennial rivers

Small perennial rivers tend to be only of local or regional importance because of their much more limited discharge. Like the larger rivers, they tend to be exogenous. In the Middle East, the River Yaramuk and the River Zarqa, both tributaries of the Jordan are examples of smaller perennial rivers. In the Atacama there are few permanent streams, but where they do occur they flow down from the Andes where they are fed by meltwater from the snowy peaks and glaciers. The Rios Lluta and Loa are two examples of small permanent rivers in northern Chile.

c) Wadis

There are thousands of dry river beds throughout the arid and semi-arid regions of the world. Most of these have their sources within the desert environment, but occasionally some of the larger wadis are **allogenic**, i.e. they have their origins outside the desert environment and bring some water into it, until a point is reached where evaporation exceeds discharge and they dry up. Wadi channels tend to flow with water only after the sporadic (but often heavy) rainstorms associated with desert areas. Much of the runoff from flash floods along wadi bottoms is quickly lost through evaporation and infiltration unless there are mechanisms in place to trap it. Throughout desert areas catchment dams have been put across wadis to trap runoff. In Saudi Arabia there are more than 60 new dams being created across major wadis. In the Australian Outback many of the extensive cattle ranches have water catchment dams built over dry river beds to supplement their groundwater supplies. In the semi-arid badlands around Matmata in southern Tunisia, the old catchment dams across wadis (known locally as *jessours*) were replaced by much bigger and sturdier structures between 1999 and 2001.

d) Shallow aquifers

Aquifers can occur at various depths. Those closer to the surface are not only more likely to be exploited for water supply and irrigation, but are also more likely to be topped up by the local rainfall. Many desert areas have limestone hills or mountains within them or close to them which can be exploited as groundwater resources. Parts of the Atlas mountains in North Africa, the Lebanon and Anti-Lebanon ranges in Syria, Israel and Lebanon, the Djebel Akhdar in Oman and parts of the Sierra Nevada and neighbouring ranges in South West USA have shallow aquifers which are exploited for their water supplies. Over-exploitation can lead to the exceeding of the **practical sustained yield**, i.e. more water is taken out than can be replenished. Along the coastlines of certain countries, such as Libya and Israel, population growth has led to over-exploitation which has caused a lowering of the water table and the subsequent contamination of the groundwater with saltwater from the sea.

e) Deep fossil water aquifers

Deep water aquifers are much more abundant in many arid areas of the world than shallow aquifers. They are great underground reservoirs of water which cover thousands of square kilometres. Almost half of the Arabian Peninsula is underlain by the Riyadh and Rub al-Khali deep aquifer basins, and within the Sahara there are numerous deep aquifers including one beneath the Kufra-Western Desert which covers about 60 per cent of Egypt and also extends under parts of Libya and the Sudan. In Australia, the Great Artesian Basin is the biggest such structure in the world, occupying some 1 770 000 km² and underlying about 75 per cent of the Simpson Desert.

These resources should not be regarded as infinite or replenishable in the short term. These aquifers contain **fossil water** which dates back to more pluvial climatic periods in the past. In the case of the Sahara and Arabian deserts, much of the water dates back to the end of the last Ice Age, 10 000 years ago; the water in the Great Artesian Basin of Australia is believed to date back 20 000 years. Rates of extraction from these aquifers generally exceed rates of refill, especially as the present rainfall is so much lower than in the past.

f) Direct interception of coastal mist and fog

It is estimated that in the hyper-arid locations which have regular mist and fog, these sources of condensed water vapour could account for the equivalent of 30–60 mm of rainfall. If this moisture could be exploited it would effectively supplement local meteorological water supplies for some communities by up to 500 per cent. In the Atacama, hillsides around coastal settlements such as Arica and Iquique are spread with very fine plastic meshing which is angled in such a way to funnel intercepted fog moisture into pipelines and storage tanks. This technology was pioneered in 1992 with Canadian funding in the fishing village of Chungungo in Chile, which became the first place to be self-sufficient from fog water.

g) Desalination plants

A long term solution to the water shortage in arid areas is the desalination of seawater. This is a very expensive process and is only operated on a large scale in wealthier countries. In the Middle East and North Africa, for example, the biggest concentrations of desalination plants are in the oil-rich countries; Saudi Arabia, the UAE, Kuwait and Libya. Solar energy is abundant in arid areas and could in the future prove to be the best source of energy for desalination, once the technology has been developed.

h) Snow and ice

Meltwater-fed streams have already been mentioned as important sources of water in arid areas which are close to high mountain ranges e.g. the Atacama and the Gobi. In the case of the Atacama, much of the potential meltwater from the Andes is lost by evaporation before it reaches the mountain streams. The climate is so arid that much of the snow and ice on the peaks above 5000 m evaporates as soon as it melts, leaving behind strange white pillars of ice known locally as 'penitential snows' because they resemble nuns at prayer. This helps to explain why there are so few perennial streams in northern Chile, and why those that do exist have such a low discharge.

In the 1970s when Saudi Arabia was rapidly expanding economically because of the booming oil industry, all sorts of avenues were explored in an attempt to solve its water deficit. One of the most extraordinary solutions, which was taken very seriously at the time, was to build large ships which would tow icebergs from the Antarctic round to the Gulf where they would be melted to produce a supply of fresh water.

4 Drainage patterns in arid environments

Although there is little surface water in deserts, other than that from perennial rivers or during flash floods, large areas of desert surfaces are marked with patterns left behind by running water; in fact the variety of drainage patterns is almost as great as in more humid landscapes (see Figure 3.4). Larger wadis have vast floodplains with heavily braided sand and gravel beds which change form each time they run with water. Sudden changes in slope have a big impact on the way materials are deposited by wadis, as can be seen where alluvial fans and internal deltas form. Where desert ranges are close to the sea, the pattern etched out by the downcutting of wadis closely resembles the parallel patterns found in similar locations in humid areas. Where there are large, isolated mountain ranges such as the Tibesti and Hoggar ranges in the Sahara, there are well-defined radial patterns of wadis draining away from them.

The only arid areas to be totally devoid of evidence of drainage patterns are the great sand seas. Here the high porosity of the material means that rainwater quickly infiltrates and is lost underground.

The most unique aspect of arid land river patterns is that found in internal drainage basins. Less than 50 per cent of desert surface runoff drains into wadis that flow towards the seas and oceans. The high evaporation rates in desert interiors enable the drainage pattern to be internal, draining inland rather than to the sea; another name for this phenomenon is **endoreic** drainage. The Great Salt Lake in Utah, USA, Lake Eyre in Australia, Lake Chad on the southern edge of the Sahara and the Chott el Djerid in Tunisia are all examples of salt lakes located at the centre of endoreic drainage systems.

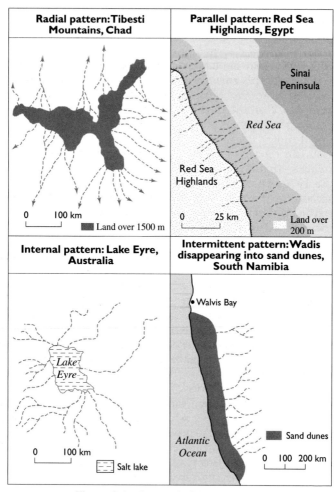

Figure 3.4 Desert drainage patterns

5 Water problems in arid environments

Many Middle Eastern and North African countries have numerous economic and political problems, resulting from high population growth, limited resources and an uneven distribution of wealth. Since the end of the Second World War, there have been numerous conflicts between nations such as the various Arab–Israeli wars and the two Gulf Wars. These conflicts have been over territory, rights of access to waterways and oil supplies. There is now increasing friction over water resources, and many experts in the region have predicted that in the future countries may well go to war over water supplies.

CASE STUDY: THE TIGRIS–EUPHRATES BASIN

The two rivers which gave birth to the ancient civilisations of Mesopotamia rise in the mountains of Anatolia in eastern Turkey. The Euphrates flows through Syria into Iraq and the Tigris flows from Turkey into Iraq (Figure 3.5), and they join to become the Shatt-el-Arab before flowing into the Gulf. The changing use of these waters by these unfriendly neighbours has led to great tension in the region. Until the 1970s, there were no large scale dams on either river, merely small barrages for small-scale local irrigation projects. Between 1968 and 1975 Syria's Tabqa Dam was constructed, ponding up 40 000 million m^3 of water in Lake Assad. This added 600 000 hectares to the country's farmland, creating a new agricultural region close to the Turkish border. The project was carried out with consent from the Iraqi government.

In the early 1970s, Turkey embarked on its South Eastern Anatolian Project (GAP) which involved 12 separate schemes on the Tigris and Euphrates. When completed in 2005, the project will irrigate 1.6 million hectares of land, thereby transforming the country's poorest region. Various small dams were inaugurated in the 1970s and 1980s (e.g. the Keban Dam), but the largest, the Atatürk Dam, was only opened in 1995. No agreement had to be made with Syria, as Turkey financed the project itself. The Atatürk Dam, named after the founder of modern Turkey, is regarded as a prestigious engineering achievement of great national pride to the Turks. To environmentalists it is just another 'mega-dam' project which will create yet more ecological problems. The Keban Dam is already silting up more rapidly than had been envisaged, but this is hardly surprising, given the rapid rates of gulley erosion on the semi-arid hillsides of eastern Anatolia. The construction of the Atatürk Dam is also leading to the drowning of several important archaeological sites. However, the political problems with Syria and Iraq are likely to be the worst legacy of the GAP project.

The Euphrates' discharge over the border into Syria was originally 30 billion m^3 per year, but the level of Lake Assad has dropped significantly as a result of the GAP project, and Syrian villagers are finding the quality of irrigation water much poorer, which is affecting both their crop yields and their health. Under pressure, Turkey has agreed to release an extra 500 cumecs of water through the Euphrates, although this was a trade-off with the Syrians in order to stop them from harbouring Kurdish PKK guerillas, who are fighting for independence from Turkey.

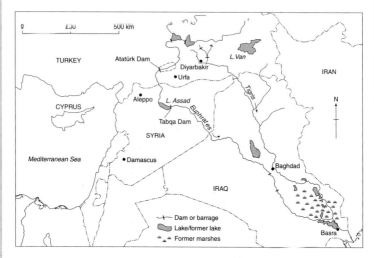

Figure 3.5 The Tigris and Euphrates Basins

Further downstream, over-irrigation and poor water quality around Basra in southern Iraq have led to high salinity levels in the soils; when the GAP scheme is finished, the situation is likely to be far worse. One of Iraq's solutions to this problem was to build its 'Third River' otherwise known as the 'Saddam River', a 565 km canal between Baghdad and Basra, designed to carry fresh water into areas with high soil salinity. This scheme is, however, full of political motives and is destroying the unique ecosystem of the Iraqi marshes and the culture of their inhabitants, the *Ma'dan* or Marsh Arabs. These marsh dwellers are independently minded Shi'ite Muslims who oppose Saddam Hussein's regime, but the waterway's construction is displacing some 50 000 *Ma'dan* threatening once and for all to destroy a unique culture and way of life.

CASE STUDY: THE JORDAN BASIN

The water situation along the eastern Mediterranean coastlands is made extremely complicated by the political situation. Although some of the higher parts of this region have a rainfall well above that of arid regions (over 1000 mm in the Lebanon range and around 600–800 mm in the hills of Galilee), most of

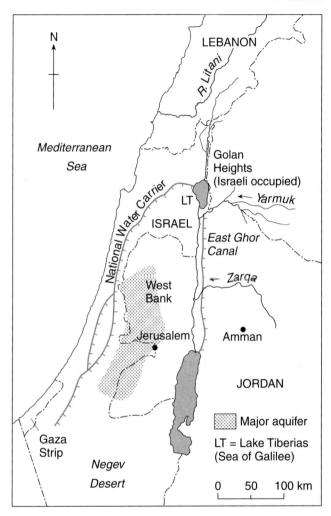

Figure 3.6 The Jordan Basin

it is arid or semi-arid and has a much denser population to support than other parts of the Middle East.

The River Jordan, the largest surface water resource in the area, has been described as a 'muddy, brackish, unimpressive river'; its average discharge at the King Hussein Bridge on the frontier between Jordan and Israel is a mere 32.5 cumecs (compared with 2650 cumecs in the Nile at Aswan). The Jordan is fed by streams from the highlands surrounding its headwaters and by some larger tributary rivers such as the Yaramuk. The main conflict in the region is between Israel and its Arab neighbours, particularly Jordan, and within the current confines of Israel, between Israelis and Palestinians. Following the foundation of the state of Israel in 1948 and the subsequent heavy immigration, the comprehensive development of water resources was a top priority; new supplies were needed to support rapid urban growth and to increase agricultural output, especially in the Negev Desert in the more arid south.

In 1951, Lake Hula and its surrounding marshes were drained and Israel started to build its National Water Carrier, to take water from the Sea of Galilee in the north to the drier lands of the south (see Figure 3.6). This brought Israel into a dispute with Jordan, which relies on water from the same river, but extracts it further downstream; this was one of the causes of the Six Days War in 1967. One of the consequences of that war was the Israeli occupation of the Syrian Golan Heights, which secured more of the River Jordan's headwater supplies for Israel. Israel also occupied the Palestinian West Bank with its abundant supplies of groundwater.

Jordan responded to Israel's National Water Carrier by constructing a parallel project, the East Ghor Canal. This has, however, put Jordan into conflict with Syria over the use of the waters of the Yaramuk and its tributaries. Currently, Israel takes 570 million m^3 of water from the River Jordan and 850 million m^3 from groundwater sources. The aquifers along the coastal belt are being over-exploited and this is causing considerable seawater seepage into the water table. Further inland are much more abundant groundwater resources under the hills of Samaria and Judea in the West Bank. It is estimated that Israel consumes 80 per cent of the water taken from these aquifers. This is not surprising when the average consumption figures are compared: Israelis use 300 litres of water per person per day (on a par with Europeans), whereas Palestinians and Jordanians use only 80 litres per person per day. As the Israelis increase the number of settlements and farms in the Palestinian territories of the West Bank, water demands will increase further. The Palestinian *intifada*, its civil disobedience uprising, has its roots partly in the Israeli handling of the water issue.

CASE STUDY: THE COLORADO BASIN, SOUTH WEST USA

As the largest perennial river in the South West USA, the Colorado River is the most heavily used source of irrigation water in the USA. Covering 94 000 km², the basin of the Colorado and its tributaries is contained within seven states, from Wyoming in the north through Utah, Colorado and Arizona, and taking in small parts of California, Nevada and New Mexico (Figure 3.7). For purposes of water extraction and management, this area has been divided into two: the Upper Basin, with a rainfall averaging over 1200 mm per annum, located high up in the Rocky Mountains, and the Lower Basin where the river flows through more arid regions and finally flows out into the Gulf of California. The Upper Basin produces 60 per cent of the Colorado's discharge, but it is within the Lower Basin that water demand for irrigation and urban supplies is much greater. 117 800 million m³ of rainfall per annum provides the main input into the Colorado River's supply system; but of this, 98 000 million m³ are lost through evaporation.

It was not until the 1890s, when this region was opened up by the westward migrations of Americans of European origin, that large-scale hydraulic engineering schemes on the Colorado were contemplated and the first major scheme was not begun until 1907. Long before this, however, the Native American Indians used the river for irrigation and built extensive networks of small-scale canals. John Wesley Powell, the Civil War veteran and explorer wrote in his 1895 book *The Exploration of the Colorado River and its Canyons*:

> 'In the valley of the Gila and on its tributaries from the northeast are the Pimas, Maricopas and Papagos. They are skilled agriculturalists, cultivating lands by irrigation. In the same region many ruined villages are found.... The people who occupied them cultivated the soil by irrigation, and their hydraulic works were on an extensive scale. They built canals scores of miles in length and built reservoirs to store water.'

The first modern scheme to be carried out in the Colorado Basin was the Salt River Project planned in 1903 and realised between 1907 and 1946. This involved the building of six dams creating a storage capacity of 2500 million m³ of water and five HEP stations. From the start the so-called 'Law of the River' was enforced; this meant that each state within the catchment area had to agree to any new scheme. In 1928 the Boulder Canyon Project which involved the building of the Hoover Dam, created

Figure 3.7 The Colorado Basin

Lake Mead with a storage capacity of almost 40 000 million m³.
At the same time the All-American Canal was constructed to take
water from Lake Mead in order to irrigate arid areas downstream
in states such as Arizona and California. This project allowed the
USA to ensure the annual delivery of 2000 million m³ of water to
Mexico. Until that time the supply at the southern end of the
river, where it enters Mexico had been highly variable from year
to year.

In the 1930s and 1940s various schemes such as the
Colorado–Big Thompson Project extended the irrigated areas
within the Upper Basin. It was found that in certain dry years
there was not enough storage capacity in the Upper Basin for
local needs, let alone those required downstream. This led to the
1956 Colorado River Storage Project which saw the construction
of several dams, the largest of which was the Glen Canyon Dam
which created Lake Powell, with a storage capacity of 30 000 million m³; HEP capacity was also greatly extended.

To make the management of the Colorado more comprehensive, in 1968 the Colorado River Basin Project (CRBP) was established. Its aim was to regulate the river flows, but also to improve navigation, reclaim schemes land through irrigation, improve water quality, maintain fish stocks and develop recreational facilities. Following on from this came the CAP (Central Arizona Project), a multi-purpose plan to provide another 400 000 hectares with irrigation and to create supplementary urban water supplies to cities such as Phoenix and Tuscon.

Over 85 per cent of the water used within the Colorado is for irrigation. In the Upper Basin 95 per cent of this comes from surface sources; in the Lower Basin almost 50 per cent comes from groundwater sources. Irrigation is in the main carried out on a low-wastage high-tech basis, using central pivot sprinkler systems and electronically controlled rapid flood techniques. The problems which are resulting from the management of the Colorado Basin are both environmental and political. There has been a noted increase in water salinity, especially further downstream, and this has caused friction between the USA and Mexico. There are also arguments between the states of the Upper Basin and those of the Lower Basin over how much water is allocated. The biggest political problems are now between farmers and the water authorities on the one hand, and the Native American Indians on the other. The indigenous people had full rights over all of the Colorado's waters long before the European Americans arrived and embarked upon any scheme to regulate the river's flow. Various court actions have taken place, many of which have been won by the Native American Indians, and many more are likely to follow.

At present, the use of the Colorado River's water has reached its limits. The highest population growth in the USA is, however, within this region, so future demands are likely to be considerably greater.

CASE STUDY: ACTION AT THE INTERNATIONAL LEVEL IN THE 'MENA' REGION

In the 1990s, the World Bank embarked on a strategy to try and avert what it identified as the 'looming water crisis' in the Middle East and North Africa (MENA) region. In 1994 it published a report entitled 'From Scarcity to Security' which laid out the facts and the statistics about water use in the region, and how a crisis might be averted in the future. The fact that the available resources within the region are severely limited can be best understood by the following figures:

- 70 per cent of the Earth's surface is covered by water
- 97 per cent of this is salt water in the oceans
- of the remaining 3 per cent, 87 per cent is locked up in ice caps, glaciers, the atmosphere, the soil or deep aquifers
- the remaining 13 per cent (i.e. 0.4 per cent of all the world's water), is readily available for human use
- of this, less than 1 per cent lies within the MENA region.

The Middle East and North Africa may not be densely populated, but have 5 per cent of the world's population surviving on less than 1 per cent of global freshwater resources. With rapid population growth (between 2.5 per cent and 3.5 per cent per annum in most of its countries), the MENA region is faced with a continuous decline in the per capita water intake; in the next 30 years this is expected to halve. The only way to avert a future crisis is to look at new ways of managing supplies and to improve water availability. To enable this, greater cooperation is needed between countries within this region in order to create partnerships and use the water more efficiently. To this end, in 1997 there was a Regional Water Conference at which the MENA Water Partnership and Action Programme was launched.

The MENA region, with a population of 284 million and a net annual renewable water supply of 355 billion m³ is the driest part of the world. The availability of fresh water is just 1280 m³ per capita per annum (about one-third of what it was in 1960), and this figure could well drop to an estimated 650 m³ per capita by 2025. Within the MENA region some countries have well below the regional average; in Gaza (Palestinian Territories) for example, the annual average available is a mere 105 m³ per capita.

Although the majority of the water is still used for irrigation in rural areas, over 60 per cent of the population in the region now live in urban areas. The region's population has a doubling time of approximately 30–35 years, but this rate is higher in urban areas which have a growth rate of over 4 per cent per annum. Already in the smaller states along the Gulf, urban water demands exceed rural ones, this pattern will soon be followed by most MENA countries. At present, 16 per cent of the regional population living in either rural areas or the poorer urban districts lack safe drinking water, and over 80 per cent lack proper sanitation.

Groundwater resources are, in some countries of the region, not being replenished as quickly as they are being used (Israel, Jordan and the Yemen are all extracting 25 per cent more than is being replenished). The situation is also becoming more critical in the Maghreb (Libya, Tunisia and Algeria). Pollution of groundwater from fertilisers, pesticides, municipal landfill sites and industrial waste is further destroying potential supplies. In Algeria, for example, the Saida and Mitidja aquifers are both threatened by agricultural and industrial pollution. There is great wastage of water in MENA countries; an estimated 30 per cent of water from flood irrigation is lost and in some urban areas there is a 50 per cent loss from evaporation, leaks and poor maintenance.

The World Bank has set out four areas of water management strategies which are being studied and developed in the MENA region.

- Increased education and public awareness campaigns. The participation of individuals in decision-making and community-based associations is regarded as important. Such participation has been carried out successfully in Tunisia for many decades, but is badly needed in many other countries in the region.
- Integration of water resource management. Each country needs to have an integrated strategy through some form of national water authority. At the same time local water authorities' work should be coordinated. Morocco and Algeria have both introduced these integrated systems.
- More efficient use of water and reduction of pollution. The high cost of water and the inefficiency of its delivery in urban areas is often due to corrupt private vendors. In rural areas the state generally has greater control over water supplies, especially for irrigation. Where reliable resources exist and the local farmers are involved in negotiating their rights there is far less inefficiency.
- New sources of water need to be identified. These could include the treatment of waste water, especially for irrigation or the desalination of brackish or sea water.

Whatever the outcome of the World Bank MENA strategy, a great deal of money is being invested in water supply in the region. The following sums have been allotted for various projects:

- Increasing water efficiency $24 billion
- Sanitation and waste water treatment $15 billion
- Conservation of water supplies $6   billion
- Environmental protection $15 billion

As a mark of success of such investments, the World Bank has set the following targets to be met by the year 2005 within the MENA region:

- a reduction of the use of irrigation water by 10 per cent, at the same time as increasing crop yields
- a reduction in water loss by 40 per cent
- a 50 per cent increase in water available for industrial and domestic use
- access to safe drinking water for 90 per cent of the population.

Summary Diagram

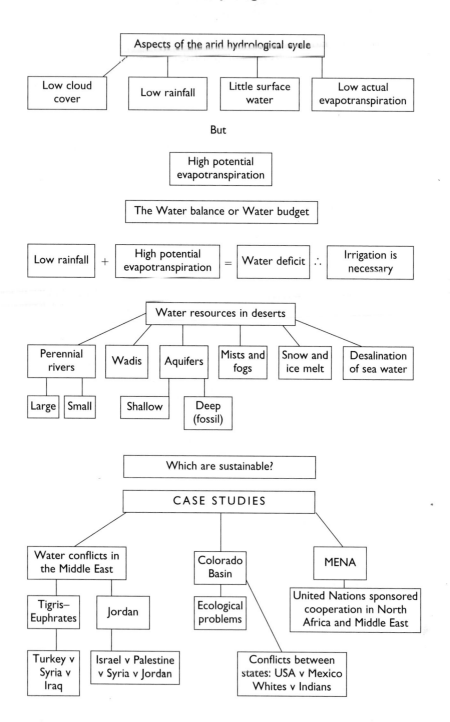

Questions

I. **a** With the aid of diagrams, describe the ways in which the hydrological cycle in arid areas differs from that in more humid areas of the world.

  b Discuss the main water resources that are available within arid areas and the degree to which they can be regarded as sustainable.

2. **a** Explain the significance of the Water budget (or balance) in the understanding of the water problems faced in arid areas.

  b 'Water is the most precious resource in arid areas.' Discuss this statement in the context of possible water shortages in the future and the way in which countries have developed their national resources.

3. **a** Draw a sketch diagram of the arid hydrological cycle, annotating on it the main ways in which it differs from a humid hydrological cycle.

  b Explain how the following water sources are important for arid areas:
  i exogenous rivers
  ii snow and ice meltwater
  iii desalination.

  c Read the case studies on the Tigris–Euphrates and Jordan Basins. Outline the main problems which may be encountered in arid regions over water supply.

4. **a** Study Figure 3.3 and explain the main differences in Water budget experienced by the four Middle Eastern cities.

  b Outline the main reasons why water supply is becoming a greater problem in arid regions than it was in the past.

  c What are some of the remedies for overcoming these problems?

4 Desert ecosystems

'Perhaps one day scientists will plan to colonise Mars. When selecting plant species to take on the mission, they will probably look for those best able to survive drought, intense light, and extreme shifts in temperature. For such a task, the earth's desert inhabitants are made to order, having been thoroughly tested in nature's laboratory for millions of years.'

Capon (1995) *Plant Survival*

1 The desert biome

The desert environment is the harshest of all the Earth's biomes, because of its limited water resources. It is wrong, however to assume that life in the desert lacks diversity. The vegetation of arid and semi-arid environments varies considerably both from one continent to another and from place to place within individual deserts. Annual precipitation totals, rainfall variability, the availability of groundwater, the depth of the water table, the salinity of the soil, the nature of the ground surface and human activity are all major influences on what will grow where within the arid zones.

The desert biome is, by its very nature, the one with the greatest percentage of plants which are adapted to aridity. There is, however, a surprising degree of biodiversity in the world's deserts. This is partly because of the way in which the main land masses were separated by

Cactus plants with their spines and pleats as adaptations to dry conditions, Central Mexico

continental drift hundreds of millions of years ago; this has allowed plant life in each desert to evolve separately over a very long period of time. Biogeographers have identified five distinctive desert and semi-desert **formations** in the world:

- The American Formation, stretching from the central Rockies of the USA, through the Mojave and Sonoran Deserts into central Mexico. In South America it continues down through the Atacama and on the other side of the Andes into Patagonia. These deserts are characterised by their huge variety of cacti.
- The Indo-Saharan Formation. This is by far the biggest of the formations, stretching across North Africa and the Horn of Africa, through the Arabian Peninsula, into Iran, Pakistan and western India. This same formation crosses the Caucasian and Himalayan mountain barriers into central Asia, the Taklamakan and Gobi Deserts. The vast area of this formation means that there is considerable regional variation within it.
- The South East Asian Formation. This is a very small arid area located mainly within Burma which has evolved its own particular associations of euphorbia and thorn trees.
- The South African Formation. Separated from the Sahara by vast tracts of savanna and rainforest, the Namib and Kalahari Deserts have their own range of unique plant species.
- The Australian Formation. Even more separated from other arid realms, the Simpson and Western Deserts have their own unique range of grasses, shrubs and bushes adapted to arid conditions.

As with desert flora, the fauna of arid regions have a wide range of adaptations to the lack of availability of water, as well as to the excessive heat

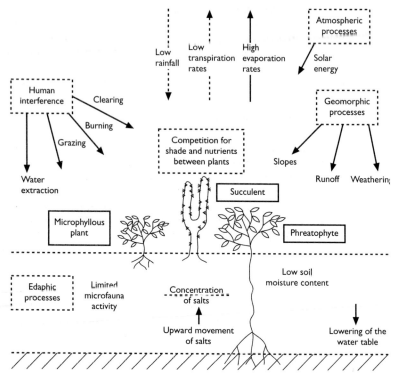

Figure 4.1 Factors influencing desert ecosystems

at certain times of the day and certain times of the year. The range of animals varies widely from one desert realm to another, reflecting long term continental separation and separate patterns of evolution.

Desert soils can be generally regarded as problem soils, the main exceptions being within the valleys of the great exogenous rivers such as the Nile, Tigris and Euphrates and in oases where long term irrigation and cultivation have greatly aided **pedogenesis**. The main environmental influences on desert soils and the way in which they develop are aridity, irregularity of rainfall, limited vegetation cover and therefore humus formation, limited soil biota and the very high evaporation rates.

The interrelationships between the various physical influences at work within arid zone ecosystems are illustrated in Figure 4.1.

2 The structure and distribution of desert vegetation

At least 75 per cent of the world's desert surfaces have some form of vegetation growing on them. Only the great erg or sand sea areas, which are constantly on the move (and therefore do not allow plants to take root on them), and the very driest places on Earth, e.g. parts of the Atacama along the northern coast of Chile, are totally devoid of vegetation.

In vegetated arid and semi-arid zones, the following characteristics are notable:

- vegetation is generally patchy rather than continuous, reflecting a low total annual rainfall
- one successfully adapted species will often form 'stands', occasionally giving way to other species
- groups of species may occur in close proximity, symbiotically relying on each other for shade
- although there is a general lack of structured layering of vegetation in deserts, there may be distinctive tree, bush and ground layers in certain favourable places such as in wadi bottoms where the water table is closer to the surface or where the semi-desert merges with a neighbouring biome such as savanna
- some species live on others as parasites
- some species have mechanisms to repel other plants to avoid competition from them
- the concept of 'climax vegetation' is not really appropriate to the desert biome, as conditions are constantly changing within the ecosystem, because of the unreliable annual rainfall.

3 Plant adaptations to aridity

Plants are adapted to arid conditions in a variety of ways; they may have physical characteristics which prevent water loss, they may store moisture in their stems or leaves, they may have deep or wide ranging root systems in order to maximise their water supply, or they may just have a short life-cycle which follows the sporadic rainstorms which are characteristic of deserts. In hyper-arid regions the plants may have to rely on coastal fogs and mists or the morning dew as their source of water supply. Some plants are adapted to saline soils as well as low rainfall, and these are some of the hardiest of all desert flora.

On their semi-arid fringes, where water supplies are more plentiful, deserts tend to merge with their neighbouring biomes. Thus, on the northern fringes of the Sahara, plant species become mixed with plants of the Mediterranean associations (e.g. the herbs and bushes of the *maquis* and *garrigue* types of scrubland). On the southern fringes of the Sahara, desert plant species take on some of the character of the dry savanna with more arboreal species such as the acacia and the baobab.

Desert vegetation can be placed into three main categories according to how it copes with the various effects of aridity.

- **Ephemerals** are plants which have a very short life-cycle. They remain for months on end in a dry, shrivelled up state loosely rooted into the desert surface. As soon as a rainstorm takes place, seeds will germinate, and the plants will flower, fruit and disperse their seeds all in a quick sequence of a few days or a few weeks. The dispersed seeds will remain dormant until the next rainstorm. Ephemerals are more common in desert and semi-desert areas where there is likely to be at least a slightly seasonal pattern to the rainfall. *Boerhavia repens* is an ephemeral common on the southern fringes of the Sahara, which has been observed to have an eight-day life-cycle. In the Mojave Desert in the USA, many of the smaller ephemerals are nicknamed 'belly plants' because of the way they appear to lie on the ground.

- **Xerophytes** are plants which in one way or another are adapted to being able to withstand drought. There are two main types of xerophytes, those that are drought tolerant and those which are drought resistant. **Succulents** are plants which not only have various mechanisms to resist drought, such as stomata which remain closed during the heat of the day in order to keep transpiration levels to a minimum, but have also developed the capacity to store water in their fleshy stems or leaves. In the Americas cacti and the different members of the agave family are the most striking examples of succulents. In Africa, Asia and Australia, the euphorbia family fill the cactus niche and there are also various types of aloe, ranging from small bushes to large trees, such as the Namibian 'quiver tree'. Transplanting between continents has altered the indigenous plant patterns; for example the prickly pear cactus is now widely seen in Africa, where it is grown for both its fruit and its effectiveness as hedging material for livestock enclosures. **Phreatophytes** are a group of plants which effectively evade drought by having long root systems which enable them to penetrate down to great depths in order to tap the natural groundwater reservoirs. They frequently inhabit wadi beds and banks where the water table is closer to the surface. The tamarisk tree can penetrate to depths of 50 m, and the mesquite to a depth of 20 m. The desert melon (*Citrullus colocynthis*) which is found in the Sahara gives the impression of being the sort of plant which could only survive in wet conditions, as it produces a large fruit full of moisture; its secret to survival is its very long tap roots which reach down to the water table.
- **Halophytes** are plants which are adapted to salty conditions. These are found on or near evaporite crusts such as salt flats. They are not exclusive to desert areas as many of the plants found on coastal salt marshes throughout the world are halophytic.

In addition to these three broad categories of plants, there are many other ways in which flora adapt to the difficult conditions experienced in arid environments. Dependency of plants on one another is widespread in deserts. **Parasitic** plants are those which have developed the ability to be opportunistic in their obtaining moisture from other plants in a one-sided relationship. In North Africa and Arabia,

Quiver trees in the Namib Desert. Succulents growing among desert 'koppies' or tors

the *orobanca* with its delicate flowers and bulbous roots is parasitic on the root systems of small succulent bushes which grow in the dunes. Lichens, which can grow in very rocky places where little else can survive, start the development of humus thereby allowing other simple plants to colonise a surface. In such a situation plants have a **symbiotic** relationship, rather than being parasitic.

Commensalism, which involves the struggle for light, space and water, is common throughout the plant kingdom, but is particularly intense within the harsh context of the arid environment. In North America various plants deter others from growing close to them in order to have exclusive access to local soil moisture by giving off toxic substances; these include the creosote bush, *Larrea mexicana* and the rubber-bearing shrub, *Parthenium argentatum.* This system of survival, known as **allelopathy**, can only be successful in areas of low rainfall, as abundant moisture in the soil would dilute the effects of any toxin.

Seed dispersal can be a problem in deserts, and many plants are adapted so that their seeds can be blown long distances by winds, or have barbed seed husks which can be readily attached to animals. The Rose of Jericho (*Anastatica hierochuntica*) is a remarkable plant which closes up tight to protect its seeds during drought periods, but opens up when it rains enabling its seeds to be dispersed by the wind.

4 The human impact on desert vegetation

Human activities can have a profound effect on the natural vegetation of arid and semi-arid regions. The main ways in which humans have an impact on the natural ecosystems of deserts are as follows:

- The complete clearance or destruction of existing vegetation to make way for settlement, roads, mines, reservoirs and other 'urbanising' or 'industrialising' projects. The major mining settlements of central Australia and northern Chile (e.g. Kalgoorlie and Chuquicamata respectively), as well as the vast areas inundated behind dams (e.g. the Nile Valley under Lake Nasser in Egypt and the Colorado Valley beneath Lake Powell) are all examples of areas where the natural ecosystems have been totally destroyed.
- The complete clearing of a natural ecosystem and its replacement by other vegetation, often of a higher density. This is seen in areas of irrigated oasis cultivation (e.g. the oases of southern Tunisia with their three layers of cropping).
- Overgrazing with animals, especially in periods of drought can lead to a great thinning of the natural vegetation. This has become a problem in many of the Sahelian countries on the southern fringe of the Sahara (e.g. Mali, Niger, Mauritania).
- Certain species are selectively collected for their economic value, making them rarer and thereby upsetting the original balance of species (e.g. the seeds of the *Moringa aptéra* are collected by the inhabitants of the Red Sea Hills in Egypt for their ben-seed oil).
- Soil erosion resulting from either excessive cultivation or extreme overgrazing can lead to the complete removal of vegetation within localised areas, especially on valley sides (e.g. around Sangha, central Mali).
- Salination of the land through overuse of irrigation systems can lead to the land being unsuitable for cultivation and when natural species regenerate they are more likely to be halophytic species (e.g. areas around Basra, southern Iraq, and the former shores of the Aral Sea in central Asia are two places badly affected by salination).
- The lowering of the water table by the over-extraction of groundwater for agricultural or other uses can lead to changes in the natural ecosystems in the surrounding areas. This is a big problem in and around the Tozeur Oasis in southern Tunisia, where hot groundwater has to be pumped up from 1.2 km underground.

5 Animal adaptations to aridity

Mobility is very important for survival for many animal species living in arid lands. Larger mammals are not generally well-adapted to arid conditions, but antelopes, gazelles and kangaroos overcome this problem by being highly mobile. These animals are able to cover the long distances between pastures and water sources rapidly. Domesticated mammals such as cattle, goats and sheep are slower moving and rely on their herders to direct them to sources of food and drink.

The camel is the best adapted large creature to be found in arid

A Tunisian boy with a desert fox or fennec

lands. As well as having the ability to store large quantities of food as fat in its hump and having a large water intake capacity, its thick-skinned lips and mouth enable it to eat thorny xerophytic plants, its long eye-lashes protect it from the effects of sandstorms and its well-padded feet enable it to cope with travel over shifting sand dunes. The Bactrian camel, which is native to central Asia, is better adapted to the cold winter desert environment than the Arabian camel or dromedary which is found in its wild and domesticated forms throughout North Africa and the Middle East. The two-humped Bactrian camel is shorter, sturdier and more hairy than its Arabian cousins enabling it to withstand the sub-zero temperatures of the winter months.

Smaller mammals are often adapted to the desert climate by being **nocturnal** and avoiding the excessive heat of the day. The desert fox or *fennec* and the desert hare of the Sahara are adapted to the heat by having large ears, which gives them a greater cooling surface. Many smaller creatures such as rats, moles and shrews burrow holes in the ground to get away from the excessive heat of the ground surface, and many reptiles and arachnids hide in the shade of rocks during the heat of the day; this avoidance of direct radiation is known as **estivation**. The outer skeletons of many insects and arachnids are almost impervious, this being a way in which they cut down on their moisture loss.

Certain lizards practise 'thermal dancing' during the heat of the day, resting alternately on just two feet at a time to avoid full contact with the hot ground surface. Similarly, some groups of insects known as **solifuges** avoid the surface heat by leaping up a few centimetres into vegetation for the odd 30 seconds to cool off; temperatures can be as much as 10–15 °C cooler just above the surface. Some lizards such as the chameleon and some snakes change their skin colour in order to reflect more heat during the day. In the foggy coastal deserts of the Atacama and Namibia many reptiles, insects and arachnids have evolved methods of utilising the moisture which is intercepted by their bodies.

6 The problem soils of arid environments

In its survey of the suitability of the world's soils for cultivation, the Food and Agriculture Organization of the United Nations (FAO), rates only 11 per cent of soils as having no limitations; 28 per cent are classified as being too dry; a further 23 per cent as having chemical problems; 22 per cent as being too shallow; 10 per cent as too wet and the remaining 6 per cent as being permafrost. The characteristics of being too dry, having chemical problems and being too shallow are all relevant concerns in the arid and semi-arid regions of the world.

The lack of moisture, the sparsity of vegetation cover and the high rates of evaporation are the three major factors which inhibit soil development in arid environments. Even where soils do form, they are generally poor, thin, lacking in humus and are highly saline. The best agricultural soils are to be found in the areas which have an abundant water supply such as the valleys of the great perennial rivers and in oases, where farmers have for centuries irrigated and culti-vated the land thereby aiding **pedogenesis**. In the general scheme for world soil classification prepared jointly by the FAO and UNESCO (United Nations Educational, Scientific and Cultural Organisation), the **zonal** soils of arid lands are called **aridisols**. Characteristically aridisols range from yellowy red to grey-brown in colour and have a very thin topsoil. Their organic content is just 1–2 per cent, which gives them a high pH value of 7.0–8.5. Under the A horizon is a zone where calcium is concentrated, and below this is a salt-rich or gypsum-rich layer which may be up to 3 m thick. Aridisols, which are very slow

CASE STUDY: ALTITUDINAL CHANGES IN PLANT AND ANIMAL LIFE IN THE ATACAMA DESERT

The Atacama is the most arid of all the world's deserts; indeed parts of it appear to have been hyper-arid for at least 3 million years. Stretching for 1000 km from north to south, it occupies a relatively narrow belt (90–150 km wide) between the Pacific coastline and the *sierras* (mountain ranges) and *altiplanos* (plateaux) of the Andes. It is only when peaks of around 5000 m are reached that there is any significant precipitation, but this falls as snow. Snowmelt is therefore one the main sources of water to be found on the western side of the Andes. Some of the altiplanos have huge *salars* or salt lakes within them; these contain evaporite minerals together with sulphurous deposits from localised volcanic activity.

Figure 4.2 is an idealised cross-section through the Atacama from the Pacific to the High Andes and shows the relationship between altitude and plant life. In the very driest places along the Pacific shore, the vegetation is at its sparsest and there are vast tracts of land totally devoid of plant life apart from lichens clinging to rock surfaces. A few isolated species of cacti and spiny bushes keep themselves alive by trapping coastal fog on their surfaces. As one moves higher up into the mountains, precipitation levels increase and so does vegetation density. At around 1000 m various microphyllous bushes, dry grasses and cacti (both the tall Echinocactus and the low 'pin-cushion' varieties) become more abundant. At around 2000 m there are great stands of a type of candalabrum cactus (*Cereus candelaris*) which has particularly large spines on its lower stem to protect it from predators and the *cabuya*, a succulent of the agave family is also common.

From around 3500 m upwards, small valleys run with meltwater-fed streams containing a type of vegetation known in Chile as *bofodel*; this is composed mainly of hardy grasses and small bushes

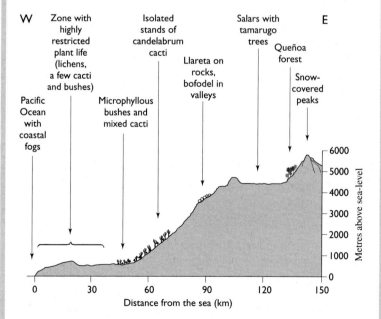

Figure 4.2 Cross-section through the Atacama from the Pacific to the Andes

which can withstand both low temperatures and high salinity. At these same altitudes the strange *llareta* or 'brain coral' can be found growing on the ground. This lumpy bright green plant looks rather like moss but under its surface of tiny tightly-packed green leaves and yellow flowers, it is rock-hard and black. The *llareta* is well adapted to the cold conditions and thrives at these altitudes because of the higher levels of UV radiation. The Incas and their descendants traditionally used the oily black resinous deposits under the living plant as fuel. At the beginning of the twentieth century the *llareta* was used to fuel the Arica–La Paz steam railway and the plant was almost 'quarried' into extinction; it is now therefore a protected species.

In the highest parts of the Andes there is some tree life. The spiny *tamarugo*, which is a type of acacia tree thrives around the edges of the *salars* where it can tap groundwater through its deep root systems which penetrate as much as 12 m into the ground. The *queñoal*, which is nicknamed 'the High Andes tree' by the Chileans, also thrives at 4000–5000 m above sea level. Small in stature, it has a very contorted trunk and branches covered by a thick bark which allows it to withstand very low temperatures, high levels of UV radiation as well as aridity.

The Atacama's altitudinal zones also influence animal life. Close to sea level, only well-adapted reptiles, insects and arachnids can cope with the hyper-aridity by having mechanisms such as those which trap the coastal mists and fogs. Up into the Andes at around the 2000 m mark the *vicuña*, a type of wild llama can be seen wandering around in small flocks in search of food. At around 3500 m these are replaced by larger flocks of the smaller but hardier *guacanos* feeding on the *bofodel* vegetation in the valleys. The members of the llama family are in fact the South American equivalents of the African and Asian camels. The other common animals found above 3000 m are the *vizcachas*, a very furry and therefore well-protected type of mountain hare.

Within the intermontane basins the *salars* contain varieties of bacteria which range in colour from emerald green to turquoise and from silver grey to pink. Deposits from localised volcanic activity have created an ideal environment for the sulphur-loving bacterium which colours the *salars* pink and provides the feeding grounds for large flocks of flamingos.

Living on the highest and driest slopes and altiplanos of the Andean Atacama are flocks of *nandu*, the Chilean rhea. These flightless birds, which can cope with a very limited water supply, are the South American equivalent of the African ostrich. The *queñoal* forest which covers the highest mountain sides provides an exclusive habitat for deer, foxes and pumas.

CASE STUDY: PLANTS AND ANIMALS IN THE NAMIB DESERT

The Namib Desert stretches 2000 km from South Africa to Angola, but the majority of it lies within Namibia itself. Like the Atacama, the Namib is a west coast desert with a cold ocean current (the Benguela) responsible for its extreme aridity along the coastline and its frequent fogs. The moisture from these fogs and mists is estimated to provide the equivalent of 30–50 mm of rainfall. The Namib has considerable variation in its scenery and, although dominated by sand dune areas immediately along the coast, it has areas of mountains, plateaux with mesas and buttes, deep canyons, and in the north, one of Africa's largest salt lakes, the Etosha Pan. It is not surprising therefore that there is a great variety of animal and plant life in the region, well adapted to the different habitats. Although the Namib has a number of strange and unique species, the majority of its wildlife has parallels in other desert areas of the world because of **convergent evolution**.

Along the coast there are various creatures that are specially adapted to using the moisture generated by the fogs and mists. The various types of **tenebrionid beetles** which inhabit the coastal dunes, actively hunt during the day but between midnight and dawn carefully position themselves with their heads down so that they can collect fog water on their backs which will then drip down towards their mouths. Some of these beetles can trap up to 40 per cent of their body weight in water, including the 'flying saucer' shaped *Lepidochora kahani*. Certain types of spiders, scorpions, snakes and lizards also have the facility to make use of fog moisture.

As in other deserts, many of the smaller creatures have adapted to the heat of the day by burrowing into the sand; these include the 'sand-diving' lizard, the meerkat, the ground squirrel and the golden mole. Certain other species evade the extreme heat of the day by occasionally leaping a few centimetres up into vegetation where temperatures are at least 10 °C cooler than at surface level; these include the solifuge spider. Some other species are physiologically adapted to the extreme heat; the bat-eared fox and the cat-like caracal both have large ears as a mechanism to cope with high temperatures.

In the less arid interior of Namibia where the landscapes are more varied and rockier, a wider range of animal life occurs, much of it similar to that of the dry savanna elsewhere in Africa. Herbivores such as gemsbok antelope, predators such as the cheetah, and scavengers such as the spotted hyena are all commonly found. The most remarkable survivor in this interior is, however, the desert elephant. It is uncertain how large this

A desert elephant in wadi bottom at Twyfelfontein, in the Namib Desert

dwindling population is today, but it is likely to be just a few hundred. These elephants are smaller than their savanna relatives and have larger padded feet which enable them to cope with walking on shifting sands.

In the north of Namibia around the Etosha Pan and the other smaller salt lakes the full range of animal life associated with the African savanna is present, despite an annual rainfall of around 200 mm and a potential evapotranspiration rate of 3000 mm. Savanna elephants, rhinos, giraffe, zebras, lions, leopards, baboons and a wide variety of antelopes are commonly found around the salt pans in the dry season.

The plant life of the Namib is equally diverse and remarkable. Various species are adapted to the lack of water in different ways; some of the species are unique to this part of the world, others have parallels in other deserts. At the most simple levels of plant life, there are highly tenacious algae and lichens. The 'window algae' grow under the translucent grains of quartz in coastal areas. The fog and dew infiltrate through the pores between the quartz grains and enough light penetrates through the quartz to allow photosynthesis to take place. Few parts of the world have such a variety of lichens as the Namib Desert. They vary greatly in colour: yellow, red, orange, grey, white, green, black, and in form: crusty, leafy or even bushy. These plants are almost always attached to rocks or gypsum crusts on the soil and rely on the fog and dew for their moisture. There are some varieties, however, which are footloose and blow in the wind until they reach hollows where they become entangled with other lichens thereby achieving a stable position. Many other types of plants use the coastal fogs and dews within the dune areas. *Trianthema hereoensis* is an example of a small succulent which inhabits coastal dunes; like the most common dune grass, *Stipagrostis sabulicola*, it has a very shallow but long, spreading root system which taps into the fog moisture layer a few centimetres below the surface.

Further inland in the rockier desert areas where there are dry water courses, there are numerous trees and bushes of a phreatophytic character, their long tap roots extending down to extract moisture from the water table. Trees found in these areas include the wild tamarisk, the false ebony and various types of acacia. One of the acacias common in Namibia is the camel thorn (*Acacia erioloba*), which not only has spines and thin leaves as a mechanism to reduce transpiration, but also secretes a thick orange resin which protects its bark from the heat and water loss.

What is most remarkable about the plant life of the Namib is the great range of succulents, many of them unique to this particular desert. Smaller types of succulents include lithops and 'mesems'. Lithops, as their name suggests, look rather like living stones; they are just a few centimetres long, grow close to the ground and intercept fog moisture as their main source of water. The 'mesem' or *Mesembryanthemum*, germinates after rainfall and has a life-cycle of just a few years; it starts off as a bright green plant but gradually goes red due to the salt content of the soil.

There is a wide range of euphorbia in the Namib. These are similar to the cactus and have the same swollen stems, spines and moisture storage facilities in them. Few are as spectacular as their American relatives, but the *Euphorbia virosa* grows to several metres in height and produces flowers similar to those of cacti. Certain trees and bushes of the Namib have bulbous swollen trunks and branches where their moisture is stored, and these resemble smaller versions of the baobab tree which is commonly found further north in the African savanna.

Another well-represented group of succulents are aloes. These have spikey, water-retaining fleshy leaves. Aloes vary in size from just a metre or so high in the case of the bright red flowered *Aloe namibiesis* to the size of a fully grown tree. The quiver tree (*Aloe dichotoma*), has a tall branchless trunk with smooth but flaking bark, and a branching crown full of succulent leaves. Although unique to the Namib, the quiver tree is similar to the 'dragon tree' of the Canary Islands, which are located at a similar latitude on the other side of the Equator. The strangest of all Namibian succulents is the Welwitschia (*Welwitschia mirabilis*). This is found inland in dry watercourses and takes its moisture from both the

> ground and fogs. This plant, which can live for hundreds of years, grows close to the ground and has long spreading ragged leaves. Scientists have found the Welwitschia difficult to categorise; its closest surviving relative appears to be the pine tree as both plants produce fruit in the form of cones.

forming soils, are found mainly on gentle slopes and they are easily eroded by water or wind.

In areas where mineral salts are more abundant, aridisols are replaced by two types of **halomorphic soils, solonchaks** and **solonetzs**. Solonchaks are found in areas with very high evaporation rates, such as within mountain basins and on salt lake beds. They form on unconsolidated material such as loess and alluvium, where the seasonal fluctuations of a high water table bring big concentrations of salts to the surface. These carbonates and chlorides form a thick crust which makes the soil infertile. Although horizons are barely distinguishable within solonchaks, the salinity decreases with depth. Solonetzs form in areas where there is enough rain to cause some leaching, but not enough to wash the minerals right out of the soil. The surface layers of the soil are not saline, but the B horizon is rich in sodium salts and very clayey, which renders the soil difficult to cultivate successfully. Figure 4.3 shows the soil profiles of aridisols, solonchaks and solonetzs.

7 The human impact on soils in arid environments

Few parts of the world are as susceptible to such a variety of forms of soil degradation as arid and semi-arid lands. Deserts, as they are marginal areas with poorly developed soils, have to be tended very carefully where they are being used for agriculture, whether it be arable land or pasture. When used for cultivation, overcropping can lead to the removal of valuable nutrients and eventual soil exhaustion. Over-irrigation can lead to the build-up of infertile evaporite layers, and the overuse of chemical fertilisers and pesticides can also leave the soil barren. Overgrazing is the biggest potential threat to the soil where desert environments are being used for pastoralism, as it causes a reduction of vegetation cover and a loss of humus in the soil. In all these cases the topsoil is left exposed and susceptible to further degradation by water or wind erosion. Topsoil removal can be particularly rapid in arid and semi-arid regions because of the low density or even a complete absence of vegetation cover.

Water erosion is the greater problem on slopes which have been denuded of vegetation for one reason or another. Falling and running water erodes in three major ways:

• **Raindrop impact,** which is particularly effective in heavy thunder-

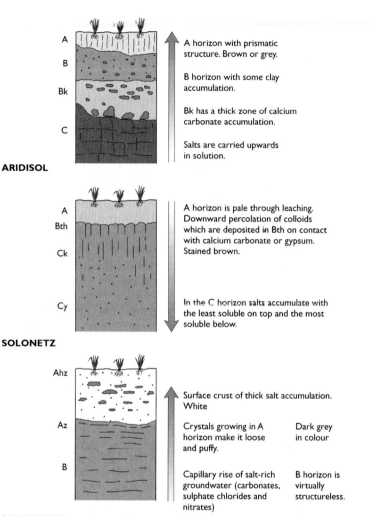

Figure 4.3 Three desert soil profiles

storms where raindrops are large in size and storm intensity is high. The resulting erosion creates pitted surfaces with an uneven removal of the topsoil.

- **Gully erosion.** Surface runoff carves out gullies and rills on slopes, taking the lines of least resistance. This also produces a selective and irregular pattern of topsoil removal.
- **Piping.** In areas which have surface cracks in unconsolidated materials, runoff may quickly infiltrate and carve out underground channels, creating warrens of soil instability.

Wind erosion is associated with flatter surfaces, especially where there are wide open spaces with few natural or human features which could act as wind breaks. Loose dry topsoil can be rapidly removed and transported long distances by strong seasonal desert winds. The large scale removal of topsoil by wind is taking place in central Asia today in the lands around the Aral Sea, causing as much ecological damage as did the great Dust Bowl disaster of the semi-arid high plains in the US Midwest in the 1930s.

It is essential for agriculturalists in desert areas to keep the natural balance in check in order to preserve soil fertility. Over-irrigation changes the chemical composition of the topsoil as a result of the upward movement of salts through evaporation, thereby increasing the salinity of the soil and effectively creating a solonchak. The over-extraction of water, which is also associated with the overuse of irrigation, leads to a lowering of the water table which in turn causes the soil to dry out and become less fertile; pedologists call this process **aridification.**

CASE STUDY: SOILS AND VEGETATION ON A TRANSECT THROUGH A MOUNTAIN BASIN IN THE SOUTH WEST USA

The **associations** of plants to be found within arid and semi-arid ecosystems vary within short distances because of changes in the quality of the soil and drainage. Factors influencing these variations include: the nature of individual landforms, the steepness of slope, the parent material upon which the soil has developed, the level of the water table and the rates of evapotranspiration.

Figure 4.4 is an idealised transect through a mountain basin in Arizona. On the higher mountain slopes the soils are very thin, rocky and dry. Very little water is available and plants growing here are predominantly varieties of succulents that can cope with long periods of drought, such as the yucca and agave. The most hardy of these succulents is the Joshua tree, a giant yucca which grows at altitudes up to 900 m and can survive both the severe winter temperatures and the very hot summers. One of its protective devices is the way in which its old leaves fold back to insulate the new shoots from climatic extremes.

On the upper bahadas (gentle slopes), where soils are made of coarse sands, gravels and pebbles and are very well drained, the water table is very deep and so, once again, succulent plants dominate. Here the very tall cacti such as the saguaro are well adapted to both the coarse soils and low level of soil moisture. The saguaro, which can grow up to 30 m in height, has pleated sides which enables it to expand rapidly and absorb water fol-

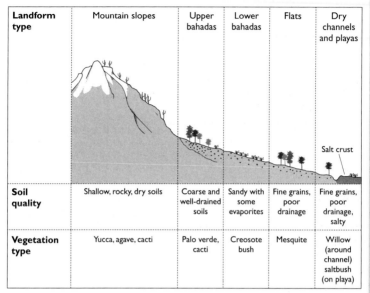

Landform type	Mountain slopes	Upper bahadas	Lower bahadas	Flats	Dry channels and playas
Soil quality	Shallow, rocky, dry soils	Coarse and well-drained soils	Sandy with some evaporites	Fine grains, poor drainage	Fine grains, poor drainage, salty
Vegetation type	Yucca, agave, cacti	Palo verde, cacti	Creosote bush	Mesquite	Willow (around channel) saltbush (on playa)

Figure 4.4   Cross-section through a mountain basin in Arizona

lowing rainfall. These pleats also create shady zones on the cactus's flanks, thereby reducing transpiration rates. The whole of the green part of the saguaro is covered in a waxy cuticle, which, as with other succulents, is the plant's first line of resistance to drought. The paloverde tree is another plant with succulent characteristics found at this level.

The lower parts of the bahadas have much finer sandy and silty soil, sometimes with a pan of evaporite material below the surface. With no surface water and a water table at some depth, root systems are of great importance to plants living here. Phreatophytes with root systems which extend right down to the water table can survive well here. At the same time, allelopathic plants such as the creosote and brittle brush which poison potential neighbours, are equally well adapted as they can maximise their intake of what soil moisture is present, without competition.

On the flats between the bahadas and the centre of the intermontane basin, the soil is extremely fine and the drainage is poor, therefore some surface water may linger following rainstorms. The soil is also quite saline. Being close to the centre of the basin and at a low elevation, this location can be extremely hot in the summer months. The mesquite and the smoke tree are particularly well adapted to these conditions. The mesquite is a short woody tree with micropyhllous leaves which, being very small, give it a very low rate of transpiration. The smoke tree is almost leafless and has a light grey bark which reflects insolation and keeps it cool; it also has spines which lower its transpiration potential.

At the centre of the basin is a playa or salt lake, which remains dry for much of the year. The high salt content of the soil determines that many, if not all of the plant species need to be halophytes, such as the sagebrush, shadskale and desert holly. The desert holly, as well as being salt tolerant has silver leaves, which gives it a high albedo, keeping it relatively cool and reducing its water loss through transpiration.

Where dry water courses cut through the lower parts of the basin there may be other associations of plants. These wadis, or washes as they are known in Arizona, have dense vegetation in their beds because the water table is closer to the surface. The desert willow and the ironwood are two species of trees which live in the bottom of washes. The ironwood relies upon washes and their occasional flash floods as a mode of seed dispersal.

CASE STUDY: THE NEFTA OASIS – AN ARTIFICAL ECOSYSTEM IN SOUTHERN TUNISIA

Around the basin of the Chott el Djerid in southern Tunisia are many oases which are based on the natural spring water supplies which gush out of the ground to form surface streams close to the edge of the basin. One of these complexes of springs feeds the town of Nefta and its large palmery. The availability of water here has supported dense vegetation for hundreds, if not thousands of years. There were a total of 152 springs in the Nefta oasis which were added to in the 1960s by artificial boreholes, in order to create a total irrigated area of 950 hectares. The natural vegetation of the area would have been a dense association of trees, bushes and, in the wettest places, marshland and reedbeds; but no written description of this exists as the oasis has been intensively cultivated for countless centuries. The oasis itself is in a dramatic setting; a bowl-shaped depression extends down the hillside from the city, in the middle of which is a sparkling blue artificial lake in which some of the springwater is collected. The water is then released out through a number of sluices feeding a network of canals which irrigate parts of the oasis. This steep-sided depression is known locally as *La Corbeille*, the French word for a bread basket. Some of the slopes of the depression are unstable and containment of this form of mass movement has been attempted. Until the mid-1980s the slopes were terraced and palm fronds were driven into the ground in rows to stabilise the terraces, with a good degree of success. In the 1980s, corrugated concrete replaced the palm fronds as it was thought to be a more durable material. By the end of the 1980s, the concrete had proved to be totally inadequate and had disintegrated in the same way that soft rocks are weathered mechanically in deserts. Mass movement had consequently accelerated. By the middle of the 1990s the more traditional palm fronds had been re-introduced as the best material for slope stabilisation.

The dense natural vegetation cover of the oasis has been replaced by a dense artificial one. The irrigated oasis gardens have traditionally been operated on a three-layered basis. The top layer is formed by the crowns of fronds of the date palm trees which need the dry heat provided by the desert climate and occasional watering. Below the palms is a middle layer of smaller trees which require some shade from the direct heat of the Sun, and this is provided by the palm tree canopies. Planted in this middle layer are tree crops which thrive in a Mediterranean-type climate, these include figs, olives, pomegranates and almonds, all of which require a moderate amount of irrigation. At ground level are vegetables and fodder crops grown in shallow earth-banked basins which are watered regularly. Crops such as onions, carrots, potatoes, beans, chilli peppers, aubergines and clover benefit from the shade provided by the two tree layers above them.

Although created by humans, the interrelationships seen within this oasis ecosystem are very similar to those which might occur naturally, where one group of plants relies upon another in order to survive under what would otherwise be difficult conditions. There are currently problems of degradation in the Nefta Oasis due to the decline in the water supply on the one hand and the increased demands being made upon the land on the other. The water table in the whole region along the shores of the Chott-el-Djerid is declining; in some places new boreholes need to be drilled down 1.5 km before water is found. In Nefta only 82 of the original 152 natural springs are still producing water. The government is drilling between 10 and 20 new boreholes a year, but some of these become exhausted after a few years; supply cannot keep up with demand. There are at present a total of 41 000 palm trees in the oasis, owned by around 1000 individual farmers, although there are only 10 main extended family groups involved. Until about 20 years ago, big, open stone-lined wells with buckets winched up by animals were the main sources of water to be used in the oasis. Now these have been replaced by government regulated wells which are operated by motor pumps feeding water into the networks of irrigation channels and pipes. The lack of rain in certain winters such as those of 1997 and 2001, together with the unreliability of any seasonal rain, has led to a strict rationing of the water supply. The government has restricted irrigation to one day a week. The length of time allowed depends on which part of the oasis a farm plot is located and what crops are being grown at ground level. The water allowance therefore varies from just 15 minutes to 1 hour 10 minutes per week. Some areas of the oasis are not surprisingly suffering from a decline in productivity.

Although the Nefta Oasis is still a striking place to visit, the deline of the local water table has taken away some of its old luxuriance and beauty which were described by writers who visited early last century. Norman Douglas wrote about Nefta in his 1912 book *Fountains in the Sand*, in which he describes countless springs flowing out of the slopes of La Corbeille (no longer visible today), and then forming 'glad pools of blue and green that mirror the foliage with impeccable truthfulness, and then after coursing in disacted filaments ... speed downhill towards the oasis.' Sacheverell Sitwell, who in his book *Mauretania* recounted his travels through North Africa in the late 1930s, also found Nefta in a more pristine state than today. '(Nefta) ... is a perfect microcosm of of the true Sahra. Its oasis is of tremendous fertility and, below the town the water springs, or fountains, with clear pools into which children are ready to plunge for coppers, and luxuriant palm trees, make it a beautiful Oriental vision.'

Summary Diagram

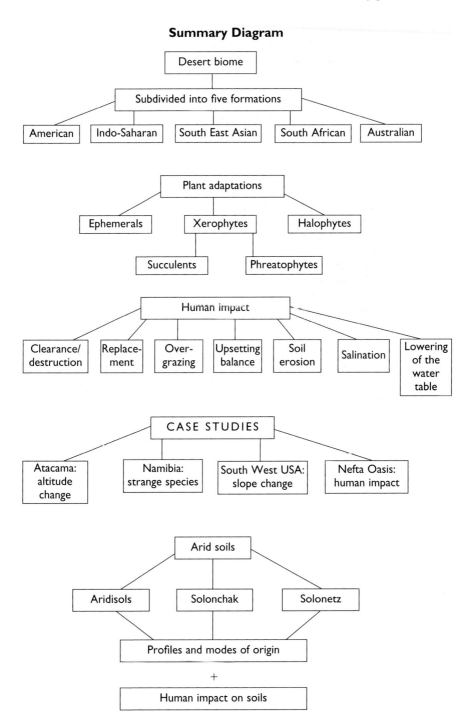

Questions

1. **a** Explain the main factors which influence what types of plant life can survive in arid environments.
    b Within arid regions there are considerable variations in wildlife according to local conditions. With reference to specific localities, explain what local variations occur in desert ecosystems.
2. **a** What are the main processes which determine the nature of soils in arid areas?
    b Discuss how the human impact on desert ecosystems can lead to both degradation and improvement of the natural environment.
3. **a** Using the information shown in Figure 4.1, explain the ways in which climatic, edaphic, human and geomorphic factors have an impact on the desert ecosystem.
    b Examine the different ways in which plants can be adapted to arid environments.
    c What are the main characteristics of desert soils?
4. **a** Using the information shown in Figures 4.2 and 4.4, outline the ways in which altitude, slope and drainage may influence what vegetation grows where within arid environments.
    b Outline the ways in which human activities can lead to the degradation and destruction of desert vegetation.
    c What are the main relationships between soils and human activity within arid areas?

5 Desert landscapes, processes and landforms

'The immense diversity of environmental conditions in deserts conspires to create a complex pattern of landforms. It is quite false to believe that lack of water engenders a simple distinctive geomorphology. Climatic variability over space and time has led to not only a large range of landform-producing processes, but also to a complementary diversity of soils and vegetation which, in turn, influences geomorphological activity.'
Goudie and Watson (1980) *Desert Geomorphology*

1 The nature of arid landscapes

Arid and semi-arid areas have a wide range of landscapes and landforms. These are a product of their underlying geological structure and the processes of weathering, wind action and water action working on the land surface through time. As the climates of deserts change over long timescales, the fundamental processes at work on the landscape also change. One of the most striking characteristics of arid landscapes is their angularity and abrupt changes in slope. This is in contrast with the more humid landscapes found in areas such as North West Europe which are generally much smoother and more rounded than arid ones. The explanation of these differences is found in the amount of vegetation cover. In deserts the overall sparsity of vegetation not only lays bare various geological structures within the landscape, but also allows geomorphic processes to take place rapidly without being impeded by the presence of plant roots. Humid landscapes by contrast are normally covered in vegetation, the roots of which slow down the rates of denudation.

Until the 1960s, the role of wind action in the creation of desert landforms was generally overstated, and the role of water was often regarded as far less important. Research carried out in deserts in the 1970s and 1980s led to a much better understanding of the importance of the work of running water in arid areas, particularly in the context of past climates when rainfall was far higher and erosional landforms were consequently being etched out at a more rapid rate. In fact the role of water was often over-exaggerated by geomorphologists at that time, and since the 1990s more credence has been given once again to the role of wind action. Similarly, there have been problems in interpreting the significance of different modes of weathering. In past decades the role of mechanical weathering was believed to be far more important than chemical action in the disintegration of rock. Detailed scientific research in the field in conjunction with experiments which simulated natural conditions in the laboratory, have led to a fuller understanding of how important chemical weathering is in deserts.

Many desert landforms are the product of **equifinality** i.e., they may have more than one possible cause of origin. As Goudie and Watson put it:

'One of the main lessons that can be gained from the study of desert landforms is that features which have a broadly similar form may have been created by processes which are fundamentally rather diverse. Simple or single explanations are not appropriate.'

2 The types of landscapes found in arid regions

Every major world desert has a great variety of scenery within it. Desert landscapes are the broad geological surfaces which may extend, over thousands of square kilometres. In contrast, the individual landforms within them may be highly varied as a result of localised changes in geological structure, climate, water availability and therefore the geomorphological processes taking place.

Numerous arid land geomorphologists have tried to classify desert landscapes; Mabbutt, in his work on the Australian deserts, has identified six major landscape units:

* desert uplands and mountains where bedrock is exposed and slopes are steep
* desert pediments which lie at the foot of upland ranges and separate them from lowlands

- stony deserts, which are plains and plateaux scattered with rock fragments
- desert rivers and floodplains, generally waterless for most of the year
- desert lake basins, including salt flats
- sand deserts.

Another similar classification is that frequently used to subdivide the Sahara and Arabian Deserts into five distinctive landscape areas; this uses many terms derived from Arabic. Although based on the North African and Middle Eastern desert areas, these types of landscapes are common to all parts of the arid world. The five categories are:

- *hamada* deserts, or bare rock surfaces such as plateaux (e.g. the Ksar Plateau in Tunisia, with its gorges and mesas)
- *reg* or stony deserts, where rock fragments are scattered over vast areas (e.g. the basalt lava flow areas of the Syrian–Jordanian Desert)
- *erg* or sandy desert, which includes huge 'sand seas' (e.g. the Grand Erg Oriental and the Grand Erg Occidental of the Sahara)
- mountain areas within deserts (e.g. the Tibesti and Hoggar ranges in the Sahara)
- intermontane basins or internal drainage basins with *chotts* (salt lakes) at

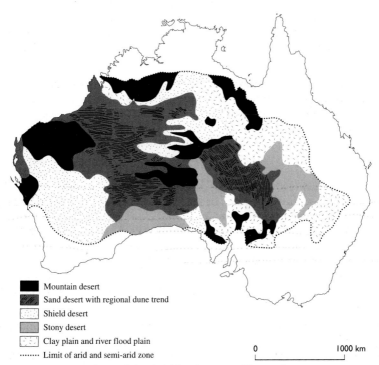

Mountain desert
Sand desert with regional dune trend
Shield desert
Stony desert
Clay plain and river flood plain
Limit of arid and semi-arid zone

0   1000 km

Figure 5.1   Arid landscapes of Australia

the centre (e.g. the Chott-el-Djerid in southern Tunisia) and other depositional drainage features such as inland deltas and wadi floodplains.

On a global scale, each of these five types of desert surfaces accounts for approximately 20 per cent of the world's total arid and semi-arid land area. Within individual deserts, the percentages are much more varied. For example, mountains account for 47 per cent of the Arabian Desert but only 16 per cent of the Australian Outback; 26 per cent of the Arabian Desert is *erg*, compared with 38 per cent of the Australian Outback.

3 Processes at work in arid landscapes

Figure 5.2 shows the main processes at work in arid areas. Weathering in all its forms tends to be highly localised, although over a long period of time it can transform large areas of desert surfaces, and prepare them for further geomorphological action. Wind action can create both large and small features within the landscape. At a large-scale, the process of deflation produces surface depressions which may be hundreds of square kilometres in extent, and the transportation of sand produces dune fields which may cover thousands of square kilometres. Abrasion, or sand-blasting, tends to create much more localised and small-scale features within deserts.

Water action is responsible for most of the large-scale features of

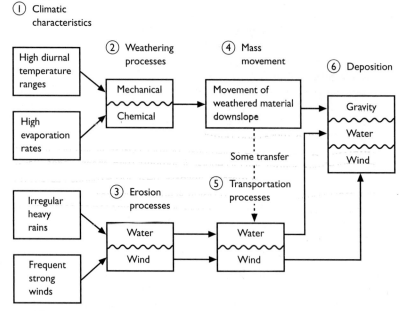

Figure 5.2 The main processes at work on landforms in arid areas

arid landscapes outside of the great sand seas. Despite the limited precipitation and run-off, the effects of running water are very dramatic in the moulding of desert landforms, by both erosion and deposition. As the present desert landscapes are the products of at least several millennia, many features would have started to form during wetter periods in the past.

4 Weathering processes in arid areas

Mechanical and chemical weathering are both widespread in desert areas; the restricted nature of vegetation cover means that biological weathering is very limited.

a) Mechanical weathering

Thermal fraction, i.e. the breaking down of rocks as a result of temperature change is the most important type of mechanical weathering in arid environments. There may be diurnal air temperature ranges as great as 25–30 °C in deserts during the summer months, and rock surface temperatures may reach as much as 75 °C. This creates a daily rhythm of expansion and contraction of rock surfaces exposed to uninterrupted solar radiation during the daylight hours. Geological structure, surface colour and chemical composition of rocks all influence the way in which they disintegrate. Rocks therefore break down in a variety of ways (Figure 5.3), the four main processes being:

- **Granular disintegration.** Grain rocks such as coarse sandstones and granite break down into grains of sand. Granite disintegrates quickly because it contains both black mica and white quartz crystals which heat up and cool down at different rates because of their colour.
- **Block separation.** Well-jointed sedimentary rocks, particularly limestone and certain types of sandstone frequently start to break down along their joints and bedding planes which are their main lines of weakness. Blocks can commonly be seen breaking off the hard cap-rocks of features such as mesas and buttes.
- **Shattering.** Some rocks which have neither coarse grains nor a blocky structure may shatter into irregular fragments with sharp edges. Basalt, which is black and highly metallic, absorbs heat and expands rapidly. It is the most common rock which disintegrates in this way.
- **Exfoliation.** The surface of any rock is more exposed to temperature change than the underlying strata. As a result some massive rocks, especially when they have a rounded profile, tend to peel off in layers. Granite and sandstone frequently disintegrate in this way.

These processes of mechanical weathering were once believed to have operated in isolation from chemical processes, and took place without the presence of water. In recent decades, research has revealed how important the presence of even a small amount of moisture is

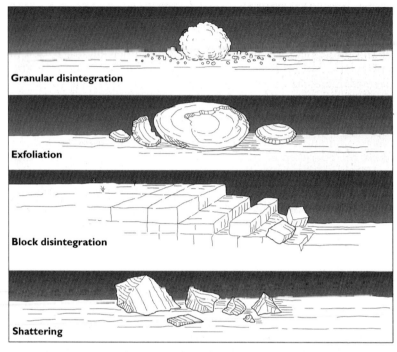

Figure 5.3 Mechanical weathering processes

within the rocks to speed up rates of mechanical weathering. Without this moisture from dew, fog or groundwater moving upwards through the rocks, mechanical weathering would almost cease to take place.

Frost shattering or **congelifraction** is a process associated with mountain areas and high latitudes. Mountain ranges within deserts, e.g. the Tibesti and Hoggar ranges in the Sahara, and the Andean regions of the Atacama, frequently experience sub-zero temperatures at night. Water, particularly from dew, which is trapped in cracks in the rocks will freeze and expand, then thaw and contract. Over time this leads to fragmentation of the rocks. Frost shattering is also widespread in cold winter deserts such as the Gobi.

The presence of water is also responsible for the breakdown of certain softer rocks in desert areas. Clays, mudstones and shales are particularly vulnerable to the process of **wetting and drying**. This occurs where spring water seeps out of the rocks, where the water table is rising, from the action of dews and fogs or during a rainy period. The soft rocks absorb water and expand, then they dry out and contract. The continuous expansion and contraction cause fragmentation. Wetting and drying can be the cause of the crumbling of mud-brick buildings in oasis villages, especially when they are located close to irrigation channels.

Exfoliation taking place on the Sinai Peninsula, Egypt

b) Chemical weathering

Mineral deposits are widespread both on desert surfaces as a result of high evaporation rates and within the groundwater due to deposition. Evaporite materials on **chotts** (salt lakes) and salt flats commonly include sodium chloride and gypsum. Coastal salt flats known in North Africa as **sebkhas** are also an important source of sodium chloride. The huge variety of other minerals to be found on the surface and in groundwater reflects the geological diversity of the local bedrock. Some minerals are more effective agents of weathering than others, either because they are more readily soluble or because their crystals grow more rapidly. Whatever the process, the presence of water is very important in ensuring that chemical weathering takes place. Even though relative humidity may be as low as 20–30 per cent in deserts, there is always some moisture present.

The main forms of chemical weathering in arid and semi-arid areas are as follows:

- **Crystal growth.** As different minerals are evaporated out of the groundwater or salts are deposited by evaporation following rainstorms their crystals grow, and this leads to rocks being prised apart until they fragment. Two properties are important here: the solubility of the mineral (so that it can get into the rocks as a solution in the first place) and the rate of crystal growth (in order to be an effective weathering agent). Sodium sulphate and magnesium sulphate are both particularly effective.
- **Hydration.** This process involves the absorption of water by an anhydrous mineral such as gypsum (calcium sulphate), which then causes expansion of the mineral and the break up of the rocks.
- **Solution.** This is the dissolving of soluble minerals in (e.g. rock salt)

water which can lead to the wearing away or disintegration of rock surfaces.

A wide range of landscape features can be created by chemical weathering. In the Atacama Desert, where thick accretions of salts can be found as a result of the region's long history of hyper-aridity, rock pillars and pedestals as well as underground tunnels can be found in the rock salt. In addition to these larger features, small scale evidence of chemical action such as flaking, crust formation and the pitting of rock surfaces can be seen in arid regions throughout the world. Two types of pitted surface forms are common: **alveoles** which are small scale hollows in a honeycombed surface and the larger scale **tafoni,** which are cave-like structures with arched overhanging entrances. Both of these features are common in the southern part of the Sinai Peninsula in Egypt.

Chemical weathering can also have an impact on artificial structures. In the central Asian republic of Uzbekistan, some of the world's most important historic Islamic buildings in the cities of Bukhara and Khiva are under threat from the salts contained in the water which is seeping out of poorly operated irrigation systems. The bricks, stones and ceramics of these centuries old World Heritage Site mosques, palaces and tombs are being attacked by efflorescences of salt which cause them to flake and crumble.

Many desert areas have accumulations of salts on or close to their surfaces. High evaporation rates and associated chemical processes help to explain their origins. Present-day arid conditions can explain some of these crusts, but others may be the result of past climates. **Gypcrete**, a type of gypsum-based crust is commonly found in hyper-arid and arid regions, whereas **calcrete**, a lime-based crust is more associated with the wetter semi-arid zones. In both cases the crusts can be formed either by the upward movement of salts already in the ground through evaporation, or by transportation by sheet run-off or wind action. Under certain conditions, a thin, hard and shiny dark red layer forms on the desert surface where iron and magnesium oxides are evaporated up through the soil; this type of deposit is known as **desert varnish**.

Laterite crusts, with a high iron content and **silcretes**, crusts with a high silica content, both of which are common in the deserts of central Australia, are more likely to have been formed in the past under more pluvial conditions as they resemble deposits which form today in humid tropical locations.

5 Wind action in arid areas (Aeolian processes)

Wind action takes three main forms in deserts: **deflation**, **abrasion** and **transportation**. All three of these require there to be loose sand and other particles so that the wind can pick them up and use them to alter the form of the landscape.

a) Deflation

Strong winds can transport huge amounts of material over vast distances and deposit them a long way from their place of origin. Deflation is the process whereby loose surface material is carried away from an area, uncovering its underlying rock structure (Figure 5.4a). Within the Sahara and particularly in Egypt, there are numerous large **deflation hollows** which are surface depressions hundreds of square kilometres in extent, created by deflation. The largest of these is the Qattara Depression, which is 134 m below sea level at its deepest point and must have involved the removal of thousands of cubic kilometres of sand and other loose material. Under certain circumstances, deflation is also responsible for the creation of stony surfaces. Where there are thick accumulations of unsorted materials, including

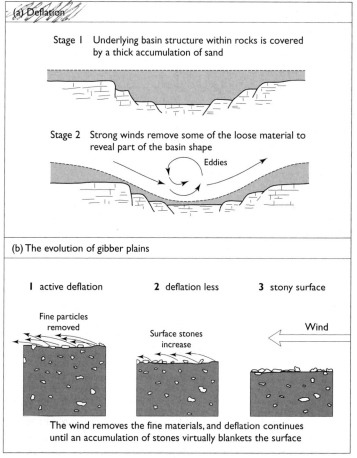

(a) Deflation

Stage 1   Underlying basin structure within rocks is covered by a thick accumulation of sand

Stage 2   Strong winds remove some of the loose material to reveal part of the basin shape

Eddies

(b) The evolution of gibber plains

1 active deflation   2 deflation less   3 stony surface

Fine particles removed

Surface stones increase

Wind

The wind removes the fine materials, and deflation continues until an accumulation of stones virtually blankets the surface

Figure 5.4   (a) Deflation (b) the evolution of gibber plains

dust, sand and stones, the wind may gradually carry the finer materials away until an almost continuous layer of stones remains on the deflated surface. (See Figure 5.4b.) The **gibber plains** of the deserts in central Australia may have been formed in this way.

b) Abrasion

Abrasion is the process of sand blasting. Sand and smaller particles are transported by the wind and driven against rock surfaces, carving them into a variety of shapes. The main factors influencing rates of abrasion are: the velocity and frequency of the wind, the wind direction, the particle sizes of the loose materials and the lithology of the rocks being eroded. At a highly localised level abrasion can create wind pitted surfaces such as **rock lattices** which may closely resemble those created by chemical weathering. Another small-scale feature to be found on desert surfaces is the **ventifact**. Ventifacts are small rocks such as pebbles, which are scattered over the ground surface and have distinctive wind-carved facets aligned with the prevailing or other winds. The number of facets depends on the nature of the winds. Where there are three distinct sides they are known as **dreikanter**.

Rock pedestals or **mushroom blocks** are also features that are often associated with abrasion; the fact that sand blasting is most effective within the first 1.5 metres from the ground surface, means that there is considerable undercutting which helps to explain the top-heavy nature of these landforms. When stratification is such that pedestal-like landforms develop as long ridges with a protective cap rock, they are known as **zeugens**.

At a larger scale within the landscape are **yardangs**. These are aerodynamically shaped ridges which often resemble the keels of upturned boats. They may be made from soft materials such as aeolianite or more resistant rocks such as limestone. It was originally believed that they only formed in soft rocks until discoveries were made in the Egyptian Western Desert of large yardangs formed in resistant limestones. Yardangs take the form of ridges that lie parallel to one another and to the prevailing winds. They are separated by troughs, and on their sides there is evidence of their mode of origin in the form of parallel wind-cut grooves. These abrasional features occur on three different scales: **mega-yardangs, meso-yardangs** and **micro-yardangs**. Mega-yardangs are impressive features which may be several kilometres long and hundreds of metres high; they are common in many parts of the Sahara where they are often made of resistant rocks, and in the Iranian Desert, where they are often fashioned from much softer mudstones and evaporite deposits. The biggest concentration of mega-yardangs, identified from air photograph interpretation, are found on the flanks of the Tibesti Mountains in the central Sahara, where they dominate 650 000 km^2 of the landscape.

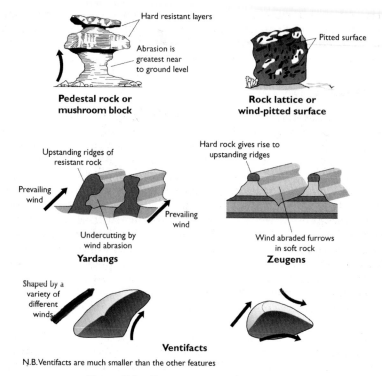

Figure 5.5 Features created by abrasion

Meso-yardangs are much smaller features, generally a few metres high and 10–15 m long. They are found widely throughout the Sahara. Micro-yardangs are just a few centimetres in height and appear as small parallel ridges of bedrock separated by small grooves scoured out by wind blown sand. (Figure 5.5 shows the main features of abrasion)

c) Transportation, dune formation and movement

Deserts are frequently very windy places. Those deserts located close to mountain ranges, especially if they are within mountain basins, can have strong seasonal localised winds. The fact that many deserts are located around latitudes 30–35° N and S means that they experience different prevailing winds in the summer months from those which blow in the winter months.

Aeolian sand covers approximately 20 per cent of the world's desert surfaces, but this varies greatly from one part of the world to another; for example, only 1 per cent of South West USA is composed of shifting sands, whereas 28 per cent of the Sahara and 38 per cent of the interior of Australia have sand dune cover. Sand dunes, unless

Meso-yardangs produced by abrasion. Fatnassa, S. Tunisia

they have become fixed, are **transportational** features. Only where dunes have become stabilised by human intervention in order to slow down encroachment upon settlements and cultivation plots, and where they have become fixed as a result of climate change, can they be regarded as depositional features. Even then, stabilised dunes such as those found on the edges of oases such as In Salah in Algeria and Dunhuang in China, can only be regarded as temporarily stable.

For sand dunes to form there must be sufficient loose material present on the desert surface. This is achieved through the long-term action of processes such as mechanical weathering and water deposition. Geomorphologists such as Bagnold recognised that sand and other loose material moves in three main ways, depending upon the size of the particles and the velocity of the wind.

- **Saltation** (from the Latin 'saltare', to leap) is the process whereby sand grains move in a series of leaps and bounds across the desert surface. Most saltation takes place just a few centimetres above the ground and is particularly associated with coarser sands and normal rather than stormy conditions. The grains will jump a short distance as they are caught by a gust of wind and then drop back to the surface.
- **Suspension** is the process where smaller particles, such as fine sand, silt and clay are often picked up by the winds and carried much further distances than in saltation. This is particularly common when high velocity winds create sandstorms. Suspended material can be carried over thousands of kilometres, e.g. material has been transported 8000 km from

West Africa to Florida in the USA, and 10 000 km from the Gobi Desert to Hawaii.

- The coarser grains of sand move along the ground by a process called **surface creep**. Creep takes place as the coarse grains of sand are rolled along the surface in a series of 'fits and starts' as and when the wind reaches the right velocity.

In the late 1980s the geomorphologist Haff suggested from his observations that a fourth transportation process was at work, namely **reptation** (from the Latin 'reptare', to move slowly). This process lies somewhere between saltation and creep and involves short distances and low level 'jumps' made by medium to coarse grained sands.

Dunes vary in size from tiny ripples on the sand surface a few millimetres high to the giant dunes found in the Gobi, Namib and Taklamakan Deserts which are between 500 and 1000 m in height. The rates of dune movement depend on wind speed and direction as well as sand particle size. Where there is a mixture of coarse and fine grains of sand, the saltation of the medium-size grains has a 'bombarding' effect on the larger grains which speeds up dune movement. The effect of the wind on surface grains causes them to shimmer and shake leading to the sorting of sand by a natural sifting process whereby smaller grains sink below the surface.

Dunes vary greatly in shape and Figure 5.6 is a classification of some of the main sand dune types. Dune shapes depend largely on the direction of prevailing and other winds; where one wind is dominant the shape is relatively simple, where there are two or more winds the shape is more complex, and when the dominant wind direction is changing, either seasonally or more permanently, the dunes also change form.

The first group of dunes are fairly small-scale and develop out of simple surface ripples where limited amounts of material are available, winds are consistent and in most cases there is some vegetation present. **Zibar** dunes are loosely formed transverse ridges up to a few metres high, which often have their movement impeded by clumps of vegetation. **Parabolic** dunes are of a similar height and are slow moving as their windward 'horns' are often partly anchored by vegetation; they are often associated with 'blow-outs' i.e. the material at the centre of them is blown away leaving a shallow depression in which a salt crust may develop. In contrast, **nebkha** dunes form on the lee side of vegetation where a small trail of sand is protected from being blown away. **Dome** dunes are simple accumulations of sand a few metres in height which are often in the process of becoming more complex forms such as **barchans**. All of these simple dune forms can be found on the edges of sand seas and where loose wind-blown sand spills over onto other desert surfaces such as salt flats.

Transverse dunes vary greatly in size, but within large areas of *erg* they can reach several hundred metres in height; as their name

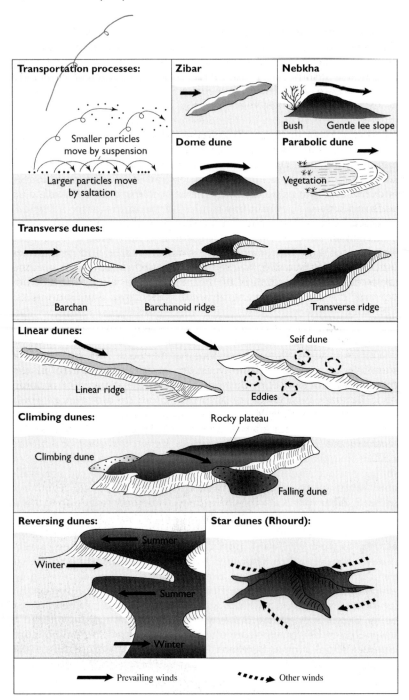

Figure 5.6 Wind transportation and sand dune type

suggests, they develop at right angles to the prevailing winds. **Barchans** are crescent shaped dunes which have a gentle windward slope and a steep leeward slope with highly mobile extending 'horns'. Saltation is responsible for the build up of the windward slope, but sand is transported down the steep face of the dune by a series of flows, slides and slumps reminiscent of larger scale mass movements in more humid landscapes. Barchans are common on the edge of the El Kharga oasis in Egypt. **Transverse ridges** have steep and gentle slopes formed in much the same way as on barchans, but have a much simpler shape. **Barchanoid** dunes are somewhere between barchans and transverse ridges in form and represent a transitional phase between the two; these dunes dominate the edge of the Dunhuang oasis in China.

Linear dunes form parallel to the prevailing wind and consequently may be hundreds of kilometres in length within the world's largest sand seas. **Linear ridges** are simple, crested elongated dunes influenced by prevailing winds, whereas **seif** dunes have serrated crests and sides as a result of the action of localised eddy winds; both types of dune are dominant in South East Libya and South West Algeria.

Reversing dunes are snaking ridges which result from an area having two well-marked seasonal winds which blow in opposing directions, and can be seen in the Sousvlei National Park dunefield of Namibia, located at a latitude where easterlies dominate in the summer and westerlies in the winter. Dunes are capable of crossing rocky barriers, even ranges of hills. When they do so they are known as **climbing** and **falling** dunes; these can be found in areas which are mainly rocky but with a small amount of shifting sand, such as the southern portion of the Sinai Peninsula, Egypt, and parts of Arizona, USA.

A reversing dune in the Namib Desert

Perhaps the most spectacular dunes are **star** dunes or **rhourds** which develop where no wind is dominant yet there are strong winds blowing in different directions. Such dunes are found over vast areas of the southern part of Saudi Arabia and in the Namib Desert around Walvis Bay. Where a sequence of star dunes form into a long serrated ridge, the resulting dune is known as a **draa**. Draas can be tens of kilometres long and can be seen in parts of the Sahara, Namib and Arabian Deserts.

6 Water action in arid areas

Water action is dominant in a large proportion of the world's desert areas, although in the vast sand seas water has little influence on geomorphological processes because of high surface permeability. Water erosion and deposition are responsible for carving out and laying down a wide range of landscape features. In highland and upland areas, such as desert mountains and plateaux, water is dominant as an erosional agent, whereas in lowland areas such as plains and intermontane basins, water deposition is more important as a depositional process.

a) Features of water erosion

Features produced by water erosion in deserts can be put into two main categories, those which are actively being carved out by moving water and are therefore increasing in size, such as wadis and canyons, and those which are left behind by the erosion process, which will decrease in size through time, such as mesas, buttes and inselbergs.

A **wadi** (from the Arabic *oued*) is a dry river bed which can be seen cutting across many different types of desert surfaces – rocky, stony, or those composed of coarse or fine alluvial deposits. Wadis may vary greatly in size, from small channels a few metres long to complex systems many hundreds of kilometres in length, such as the Wadi Malik, a dry tributary of the Nile in the Sudan or Wadi Hadramat, which forms the main strategic routeway through the mountains of southern Yemen. Although it is difficult to generalise about wadis given their great variation in size, they tend to have the following characteristics:

- their bottoms are broad and flat because of the build up of thick deposits within them
- their channels are heavily braided, also because of the thick sediments on their beds
- these braided channels change shape with each rainstorm, and change radically during heavy flash floods
- at least one if not both of the banks is steep sided, and similar to a bluff on a humid river bank; this is due to erosion rates not being impeded by vegetation cover

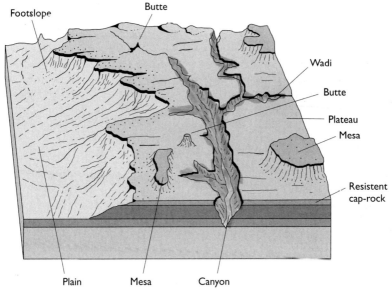

Figure 5.7 Mesa and butte landscape

- wadi banks are often deeply incised by gullies or draped with alluvial fans, both once again reflecting the lack of vegetation cover
- wadi banks are often clearly stratified with layers of sediment of different sizes. The coarser sediments such as pebbles, reflect past flash floods of considerable magnitude, whereas strata of finer sediments reflect water flow following less dramatic rainfall
- soon after a wadi has been flowing with water, a pattern of mud-cracks will appear across its bed, this is also the environment in which **armoured mud balls** form. These features which are roughly spherical and can be as large as a football are strictly speaking depositional; they have a clay core around which other layers of clay accrete and a pattern of mud cracks develops as the surface dries out.

Smaller water-carved incisions can be found throughout arid regions, these being gullies and rills similar to those found in more humid regions. They form tributaries to wadis, and are a part of the drainage network which becomes active during rainstorms when they erode headwards. The lack of vegetation cover makes them more visible within the landscape than they are in humid regions.

Canyons (desert gorges) are much deeper, steeper sided versions of wadis, and are particularly associated with the downcutting of flowing water into rock surfaces. In many cases the action of water alone cannot explain the sheer scale of a canyon, especially its depth. Over long periods of time (thousands of years), it is tectonic uplift working

Buttes in the Arches National Monument, Utah, USA

together with vertical fluvial erosion which has created the deepest desert canyons. Some of the biggest canyons are also associated with large exogenous perennial rivers which have greater erosive power than non-permanent streams. The Grand Canyon on the Colorado River in the USA, the Nile Gorge at Aswan in Egypt and the Fish River Canyon in Namibia are three examples of large-scale gorges carved by perennial rivers responding to tectonic uplift.

Mesas, buttes and **inselbergs** are all examples of relic hills which are cut out and isolated by water erosion. Mesas and buttes are typical of areas of sedimentary rocks with horizontal bedding planes and strata of varying resistances. When eroded, these rocks are left with a protective, more resistant cap-rock which is instrumental in determining their shape. Mesas are plateau-like mountains or hills which take their name from the Spanish word for a table; they tend to have steep edges delimited by canyons or wadis. Buttes are more pillar-like formations which are left behind after much more erosion has taken place and just a small upstanding relic of the more resistant rock remains. The lower flanks of mesas and buttes are commonly covered in screes which result from rockfall and other types of mass movements, prompted by mechanical weathering. Mesa and butte topography can be found in many desert areas of the world including Monument Valley, Arizona, USA, the Ksar Plateau around Tataouine in southern Tunisia and the Brandberg Plateau of Namibia. Figure 5.7 shows mesa and butte topography.

Inselberg, a word from the German meaning 'island hill', in theory covers all types of relic hills, including mesas and buttes. The term, however, is now generally taken to describe the more rounded type of isolated hill which is not only found in desert areas, but also in the more humid parts of the tropics, especially where there are crystalline rocks. There is a great deal of debate as to the conditions under which inselbergs are formed. Many geomorphologists suggest that inselbergs within arid areas are relic features not only of current erosional patterns but also of past climatic conditions when rainfall was higher. Certain types of dome-shaped granite inselbergs known as **bornhardts** are found widely throughout the savanna lands of Africa, parts of India and parts of Brazil. In these regions they may be jungle clad, but their basic shape is the same as inselbergs found in the Sahara and other deserts. Bornhardt-type inselbergs are believed to have evolved as a result of the deep chemical weathering of granite which then became exposed with the subsequent erosion of surface deposits. This type of weathering is common in more humid environments, which helps to support the theory of bornhardt evolution during wetter periods in the past. **Tors**, very similar in shape to those found on Dartmoor, can be seen perched on top of some desert inselbergs and provide further evidence of deep chemical weathering as their possible mode of origin. In the Namib and other southern African locations, the term **kopje** is often applied to a small inselberg, particularly when it has tor-like piles of rock on its summit.

Pediments are gently sloping erosional rock surfaces which are located at the footslopes of desert mountain ranges. Widely regarded by geomorphologists as the most typical feature of arid lands, their mode of origin is a matter of considerable argument and debate between desert specialists. Generally pediments have a slope angle of between 1° and 7°, and may be totally exposed, scattered with boulders and other forms of depositional debris or buried beneath features such as alluvial fans and bahadas. Many of the early theories suggested that pediments were features created by wind erosion. This would only be conceivable in a few parts of the arid realm, such as the Tibesti mountains in the central Sahara where wind action has carved out other major landscape features. Even in the Tibesti, however, the higher rainfall and radial drainage pattern would suggest that water has played an important role in pediment formation. Three of the other popular theories of pediment formation are:

- Sheet run-off coming down from desert mountains creates a smooth surface of rock without perceptible gullying. This theory can be criticised on the basis that the debris which accompanies sheet run-off is highly variable in size and therefore likely to create an uneven surface.
- Wadi streams when in flood meander and change their course rapidly, thereby causing regular rates of denudation over the whole rock surface at the mountain front over which they flow.
- Weathering and running water working together are responsible for pediment development. At the break in slope between mountain and plain, the rates of weathering are accelerated and the resulting debris is washed away within sheetflow, scouring the rock surface as it moves downslope.

Not enough is yet known about pediments and their evolution for any individual theory to hold sway. As with many other desert features, more research is needed in this field in different parts of the world in order to get a better understanding of the way in which they have evolved.

In the mid-twentieth century, King, the desert geomorphologist working in southern Africa identified pediments as being an important element within his **arid cycle of erosion**. Built on the theories of Davis, who had developed a cycle of erosion for humid temperate regions, King argued that in arid areas the landscape undergoes systematic and gradual denudation, until a state of **pediplanation** is reached. Although this theory oversimplifies both the landscapes and the processes which are at work upon them, it is still useful in explaining the overall changes that take place through time on certain desert surfaces.

b) Features of water deposition

The main features created by water deposition in deserts are **alluvial fans**, **bahadas** and **chotts** or **playas** (salt lakes and their surrounding salt flats). All of these features are commonly found within **intermontane basins**, the broad, open desert depressions with their characteristic patterns of internal drainage. Figure 5.8 shows where these features are located in relation to one another within an idealised profile of an intermontane basin. This group of features is also referred to as a **desert piedmont**. Although not themselves products of water deposition, intermontane basins have been the subject of a great deal of debate. There have been many theories about their origins, but as each theory depends very much on the local geological conditions within an individual area, it seems logical that each of the possible explanations is equally valid. The five main theories are that intermontane basins were formed by:

- deflation, where large amounts of loose material are carried away leaving a broad open depression
- the solution of limestone, which may also leave a broad open basin
- tectonic uplift, which may create two large block mountains, leaving a broad depression between them
- crustal warping, the weight of a former large body of water causing a depression to form in the Earth's crust
- the presence of a barrier, such as a sand deposit, which has blocked off an earlier drainage pattern and converted it into an endoreic one.

Alluvial fans are delta-like depositional landforms which develop where a wadi or sometimes a small perennial stream runs down from a mountain front onto a much gentler slope. When the wadi flows with water, this break in slope causes an energy loss which leads to the deposition of the vast majority of the temporary stream's load. The

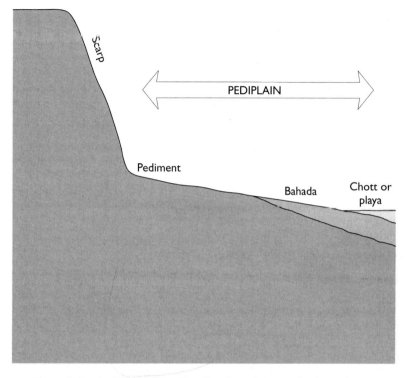

Figure 5.8 An idealised profile of an intermontane basin or desert piedmont

deposits are spread out into a fan shape by the small distributaries of the wadi, which may change their shape and direction after each rainfall event.

Alluvial fans vary considerably in size from just a few metres in length and depth to around 20 km in length and 300 m in thickness. Some of the largest examples in the world are to be found on the flanks of the mountain ranges on either side of Death Valley in California, USA. The material within these fans is graded by the intermittent running water and therefore the debris upfan is coarser than that downfan. As with the deposits on wadi beds, alluvial fan materials are stratified and cross-sections cut into them reveal layers of debris of different sizes reflecting the intensity of the rainfall events which produced them. In inhabited regions, these fans are often occupied by small farms or hamlets as their finer deposits can be developed into fertile soils, especially where the water table is close to the surface – which is generally the case close to wadi channels. The disadvantage of living in such settlements is, of course, the risk of damage from flash floods.

Many mountain ranges in arid areas are tectonically unstable and any uplift and erosion which takes place may result in wadis and their distributaries becoming incised into the very alluvial fans which they have deposited.

The coalescence of a number of alluvial fans deposited by a series of parallel wadis arriving at a mountain front leads to the formation of a **bahada** (sometimes spelt 'bajada'). These features may form the middle reaches of an intermontane basin or be downslope of any type of desert mountain range. Sometimes bahadas cover up mountain pediments, sometimes they leave them uncovered and form beyond them. They have many similar characteristics to the alluvial fans from which they form, such as the stratification and sorting of materials and the way in which they may be incised by any wadi that traverses them. There is often a distinction made between the upper and lower parts of a bahada. The lower bahada is not only made of much finer deposits than the upper part, but where it is part of an intermontane basin, it is also much more saline.

Chotts or **salt lakes** have a wide variety of names according to the arid zone of the world in which they are located. 'Chott' is the Arabic word used in North Africa, 'sabkhah' is used in the Arabian Peninsula; elsewhere, this feature is known as 'playa' in North America, 'kavir' in Iran, 'takir' in central Asia, 'salar' in South America and 'pan' in South Africa and Australia. Wherever they occur, they generally have the following common characteristics:

- they occupy low points of the desert surface
- they lack outflows to the sea
- they are occupied by ephemeral water
- they are usually very flat
- in their Water budget, evaporation greatly exceeds inputs
- they are either devoid of vegetation or host distinctive halophytic species.

Chotts may change in character from week to week or even day to day depending on when rainfall has occurred and to what extent evaporation has taken place to enable crust formation.Water which fills these lakes arrives as overland flow from higher ground.Rapid evaporation then takes place leading to the formation of a thick crust; after a long dry period the surface crust cracks up into roughly polygonal plates. Further evaporation can lead to the formation of fancifully shaped salt pillars which emerge through these cracks and then grow by accretion.

Like bahadas, chotts can be seen to have upper and lower levels. The upper levels tend only to have water on them following more extreme meteorological events, and normally remain as bare salt flats. The lower central part has water in it more frequently and therefore has a much thicker accumulation of salts. Although sodium chloride (common salt or 'halite') is often the most frequently occurring salt within chotts, there are a wide range of salts and other evaporite deposits to be found in and around them. These include calcium sulphate (gypsum), sodium sulphate, magnesium sulphate, and less commonly, potassium and magnesium chlorides. In many parts of the arid world the economic potential of evaporite material is exploited commercially. Sodium chloride is an important resource for the local economies of the Chott el Djerid in southern Tunisia, Azraq Druze in Jordan and San Pedro De Atacama in Chile; in all three of these places it is collected and transported away for processing and packaging.

Salt lakes, current and past, and the depths of salt deposits from them provide important clues for the processes of climate change and desertification (which are considered in more detail in Chapter 6).

7 Badland landscapes in semi-arid regions

Badlands are associated with semi-arid rather than arid zones. They occur where soft and relatively impermeable rocks are moulded by rapid runoff which results from heavy but irregular rainstorms. The aridity of the climate ensures insufficient vegetation cover to hold the clays, shales and other bedrock materials together, yet the rainfall is sufficiently powerful to create dramatic erosional and depositional features. Badlands are, however, not as sparsely vegetated as true deserts. As they are marginal lands, the threat of over-grazing can lead to a reduction of vegetation cover, the degradation of the land and a more rapid development of the badland landforms.

Badlands are characteristic of only small areas within deserts, e.g. 2 per cent of the Sahara, and 1 per cent of the Arabian Desert, but they do occur in scattered locations outside of the main desert belts of the world. Some examples of badlands are:

- The Badlands National Monument, South Dakota, USA. This area has an annual rainfall of 380 mm. The very name 'badlands' originated in the USA and refers to the fact that they are of little agricultural value.
- The loess plateaux of the Kansu Province of central China, an area with around 500 mm of rainfall surrounding the Huang He River.
- The area around Matmata in South West Tunisia.
- The hill-country of the Basilicata region of Italy, where the landscape bears erosion scars resulting from centuries of deforestation, and is typical of many of the drier parts of the Mediterranean area which have been mismanaged by human activity. Even though some of this hill-country has a rainfall in excess of 800 mm, it is the seasonal irregularity of the rains that puts the clays and shales at risk and gives the landscape a semi-arid character.

Figure 5.9 shows an idealised badlands landscape with its characteristic features. Drainage density is much higher than in many other

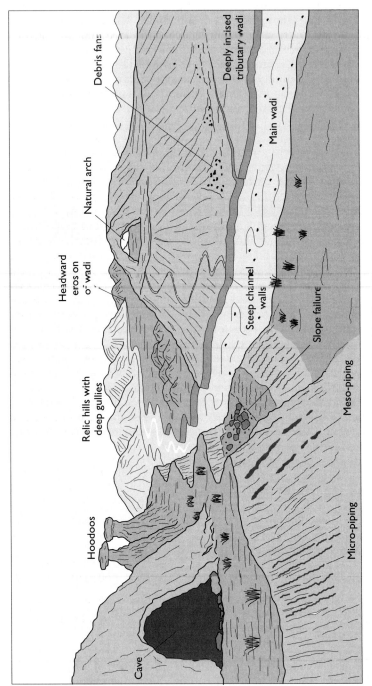

Figure 5.9 An idealised badlands landscape

landscape types because of the readily eroded bedrock material such as clays, the lack of vegetation cover and the amount of rainfall (generally between 300 and 600 mm). Also, many of the world's badlands occur in geologically young areas where tectonic uplift is still taking place, adding to the landscape's instability.

Where cracks occur on the surface of normally impermeable clays and shales, water is able to pass underground and carve out **pipes** (eroded passageways) through which water from subsequent rainstorms can then run. Piping can be identified on two scales: **micro-piping** and **meso-piping**. Micro-piping has a diameter of a few millimetres and gives surfaces a pitted or honeycombed appearance. Meso-piping can be from a few centimetres to a few metres in diameter, and where there is tunnel roof collapse, they form discontinuous gullies. Meso-piping can also produce caves and natural arches within badland landscapes, and the collapse of these structures helps to account for the large amount of mass movement in badlands.

Where areas of weakness occur within the interbedded soft rocks, particularly when surface runoff is directed towards them each time it rains, much larger and more open types of piping may occur in the form of caves. Over time these caves may be eroded back so far that they develop into natural arches. These arches may have considerable amounts of water running through them following heavy rainfall and indeed may be part of a wadi's course.

Badland wadis are generally steep sided with debris-strewn bottoms, and there is normally evidence of vertical and headward erosion within them. Tributary gullies are also very active and erode headwards cutting into hillsides and undermining them, contributing to slope collapse. Differential geology, particularly where some strata are permeable and others impermeable, contribute to slope instability. Slope failure and the slumping of hillsides are very common.

As badlands become more and more dissected by water action, relic hills develop. These hills vary considerably in size, from hundreds of metres in height in upland areas, which can be regarded as 'young' badlands landscapes, to just a few metres in height in lower lying areas which have experienced a much longer period of denudation. Generally these relic hills are rounded and heavily dissected by rills, resembling inselbergs in more arid regions. In Basilicata, southern Italy these are known locally as *dorsi di elefanti* (elephants' backs). Sometimes, where there are protective caprocks or clumps of vegetation, relic hills may take the form of pillar-like structures, which are known as **hoodoos**.

CASE STUDY: THE LANDFORMS OF SOUTHERN TUNISIA

Within the comparatively small area of southern Tunisia (approximately 80 000 km^2), a wide range of desert landscapes and landforms can be found. Figure 5.10 shows their locations.

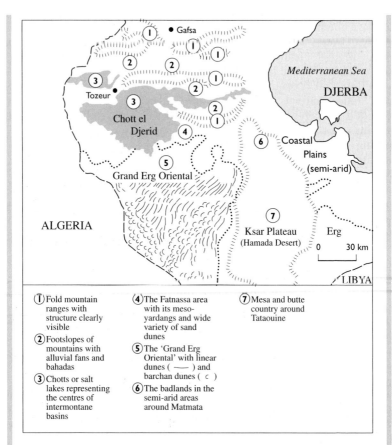

Figure 5.10   The landforms of southern Tunisia

The north of the region is bounded by several ranges of fold mountains e.g. the Djebel Tabega and the Djebel el Asker, which are outliers of the High Atlas of northern Tunisia. They are young fold mountains dating from the Tertiary orogenesis and are therefore still unstable and subject to tectonic changes which bring frequent minor Earth tremors to the area around Gafsa. Like other desert ranges made of sedimentary rocks, their strata are clearly visible because they lack vegetation cover. Where horizontally bedded they are dissected by wadis which separate their tops into numerous mesas and buttes; elsewhere they are buckled into fanciful shapes. Typically the lower slopes are covered by a thick accumulation of debris washed down during flash floods. Individual wadis have deposited alluvial fans which have subsequently merged to become bahadas. The bahada along the northern footslope of the Djebel Tabega is particularly dramatic, and having been tectonically uplifted is now deeply incised by many of the wadis which created it. The fold mountains of the north lead down to the Chott el Djerid, North Africa's largest salt lake which occupies the centre of an immense intermontane basin, some 120 km from east to west and around 60 km from north to south. The Chott receives both overland flow and fine sediments from all the wadis which surround it when it rains, but the nature of its surface is constantly changing. After a long period of intense heat and no rain it has a thick, crispy, snow-white crust; by contrast, after a heavy downpour, it can have a dull, muddy surface with large expanses of water. In the autumn and winter the Chott can be treacherous for travellers who stray from the main dry causeway which crosses it linking the oasis towns of Tozeur and Kebili.

In the south east corner of the Chott el Djerid, the sands of the Grand Erg Oriental are encroaching onto the surface of the great salt lake. Here there are thousands of small elliptical nebkha dunes, all orientated from south west to north east in alignment with the local *chehili* winds. Each nebkha dune is in the shadow of a halophytic clump of vegetation sprouting out of the salty mudflats of the Chott, and is the same pale grey colour as these mudflats.

The natural recycling of sand can be seen very clearly near the village of Fatnassa, also on the southern perimeter of the Chott el Djerid. On the outskirts of Fatnassa are some remarkable meso-yardangs made from a very soft golden sandstone. These yardangs were originally sand dunes which became compacted into rocks, and the evidence for this can be seen in their rippled surfaces. The desert winds have sculpted this aeolianite sandstone into dozens of parrallel ridges and mounds which are around two metres in height. Highly friable, these yardangs are being rapidly eroded back into sand.

The Grand Erg Oriental takes up most of the far south of Tunisia. This Erg covers a total of 192 000 km², and is the Sahara's largest area of shifting sands. Only 10 per cent of the Grand Erg Oriental lies within Tunisia, yet it contains a wide range of sand dune types. Linear and seif dunes are dominant in the northern and central parts of the Erg, whereas barchans and barchanoid dunes dominate the southern part of the region. Prevailing and secondary winds define sand dune shape. In the north of the Erg the south westerlies have a strong influence in the winter when they are the prevailing winds over the Mediterranean region and have created a marked south west to north east trend within the dune fields. Further south, where the influence of the south westerlies is weaker or non-existent, the crescent-shaped barchans largely replace the linear dunes but they are nevertheless also advancing eastwards.

Much of the eastern part of South Tunisia is taken up by the Ksar Plateau. This is an area of *hamada* or rocky desert, composed of sedimentary rocks, mainly sandstones and limestones. In the plateau country the stratification of the rocks is clearly visible, due to the sparsity of the vegetation. Over much of the area the stratification is horizontal, giving rise to a landscape dominated by gorges, mesas and buttes. As the plateau is high up and close to the Mediterranean Sea, it has a higher rainfall than further inland; this means that rates of erosion are greater, which is evident from the depth of some of the gorges.

The Tozeur Oasis provides evidence of the importance of past tectonic and climatic changes in the explanation of present desert landscapes. On the edge of the oasis is a series of small inselbergs about 20 m in height and made mainly of soft mudstones and clays. From the tops of these there is a clear view of the Chott el Djerid which represents the current base level of erosion. The inselberg tops are approximately the same height and are merely relic hills which represent all that is left of a former surface, itself once the base level of erosion. Beneath the inselbergs is a deeply incised wadi, the creation of which can only be explained in terms of tectonic uplift; the flat bottom of this gorge-like wadi is the future base level of erosion.

Not only is there this evidence of tectonic change at Tozeur, but on close inspection, the inselbergs have small layers of gypsum, the evaporite material, interbedded within the mudstones and clays. The mudstones and clays were laid down during past pluvial periods, when the Chott el Djerid was much larger and was an inlet of the Mediterranean. In contrast, the gypsum was laid down in much drier periods, when the Chott was, as it is now, a vast salt lake not connected to the sea where evaporation is the dominant physical process.

In the area around the small town of Matmata, there are thick deposits of fine wind deposited loess which have been loosely 'cemented' together by the presence of gypsum. The Matmata region can be classified as semi-arid because it has a higher rainfall than other parts of southern Tunisia (around 350 mm per annum). In common with other semi-arid parts of the world which have relatively soft rocks, the landscapes around Matmata have all the typical features of badlands. There are heavily gullied hummocky hills, highly braided main wadi channels, piping on various scales which in places has created natural arches, landslides, slumps and the scars left behind by mass movement. It is an eerie landscape which has lent itself well as the setting for various science fiction films. The area is pitted with troglodyte dwellings carved out of the cemented loess by the indigenous Berber population, hundreds of years ago.

Summary Diagram

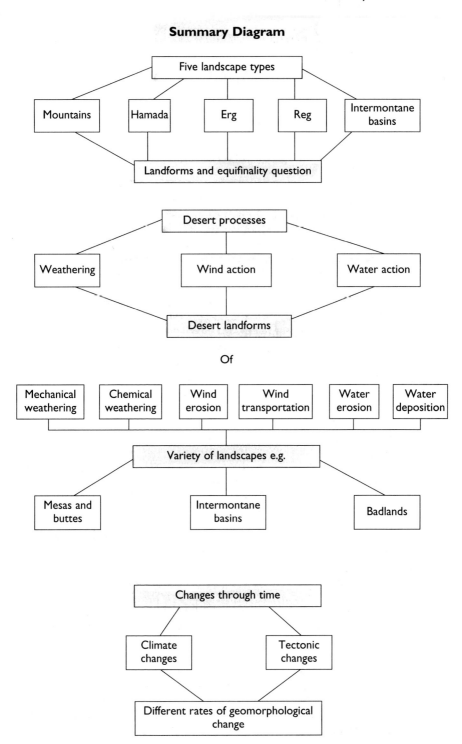

Questions

1. a Explain the physical conditions which create the different types of mechanical and chemical weathering processes which take place in arid environments.

   b Assess the relative importance of wind and water action in the formation of desert landscapes and landforms.

2. a With the aid of diagrams explain the evolution of the various features which occur within intermontane basins in deserts and how they may change over time.

   b Explain why many present day features within desert landscapes are believed to be the product of environmental conditions of the past.

3. a With reference to Figure 5.2, explain how the different geomorphic processes at work in arid areas relate to aspects of desert climates.

   b Describe how the main types of mechanical weathering in deserts take place in relation to both climate and rock type.

   c Explain why water action can be regarded as more important than wind action in certain desert areas.

4. a Account for the wide variety of sand dune types to be found in desert areas. Sketch diagrams will be useful to illustrate your answer.

   b Choose three features from Figure 5.7 and with the aid of sketch diagrams explain their mode of origin.

   c Why are tectonic changes a major factor in the evolution of such a landscape as in Figure 5.7?

6 Desertification and climate change

'Arid areas are essentially lands of high risk and of uncertainty ... but the precise kind of risk is immensely diversified, the range of uncertainty is great, and each sector of the zone provides its own special problems and opportunities for human occupation.'

Hills (1966) *Arid Lands*

1 The use of the term desertification

Desertification is a term which, in its popularised form, especially when used by the media, conjures up the image of the world's deserts expanding rapidly and irreversibly to engulf the productive farmland on its margins. The two periods of drought in the Sahel on the southern edges of the Sahara in 1968–74 and 1979–84, and the famine and human disruption they caused, brought the world's attention to the way in which deserts could expand. At the time when these droughts occurred, a great deal was written about the Sahel by experts and non-experts alike; television screens and newspapers were filled with images of parched lands, dying cattle and emaciated people. As a consequence of all this, the concept of what constitutes desertification has become confused, and very often its effects have been greatly over-exaggerated.

By the late 1980s, remote sensing was becoming a much more exact science and the interpretation of satellite images more accurate. New images were showing that deserts tend to expand and contract according to alternating drier and wetter periods. The climatic events in Africa during the early 1990s supported this idea. While the Sahel had abundant water in is rainy seasons, parts of the central/southern African savanna region, notably in Zambia and Zimbabwe, suffered from drought as their rainy seasons failed to arrive for a few years running. Satellite images taken at this time showed the Sahara retreating back northwards, while the deserts of southern Africa were starting to encroach on the neighbouring dry savanna lands.

However, severe land degradation, which can lead to desertification is clearly taking place. Concern with the state of the land in over 100 countries led to the establishment in 1996 of the United Nations Convention to Combat Desertification (UNCCD), one of many organisations devoted to the monitoring of desertification and finding remedies for it. By the year 2000, 32 member states had drawn up action plans under the supervision of the UNCCD; these plans involve local people at the 'grass roots' level and therefore the expertise of traditional farmers is taken into account alongside that of internationally respected agricultural scientists.

2 What is desertification?

One of the greatest problems concerning desertification is giving the term an exact definition. Many different writers have come up with different definitions. In the first place desertification must not be confused with **drought**. Drought is a short term period in which the climate is drier than normal, when seasonal rains fail to arrive, but from which the natural environment can recover quickly.

There is considerable confusion in the use of the terms **land degradation** and desertification; they are sometimes regarded as interchangeable, which they are not. Land degradation is the process whereby the soil becomes less productive as a result of physical factors such as drought or human factors connected with bad management of the land such as overgrazing. **Desertification** results from large-scale and long term land degradation, specifically in the drier parts of the world.

In 1978 at the UN Conference on Desertification (UNCOD) a simple definition was used:

'Desertification is the diminution or destruction of the biological potential of the land which can lead ultimately to desert-like conditions.'

This definition does not put the cause of desertification as being either human or natural, it is left open for interpretation. Neither does the definition include the world 'irreversible', but merely states that change takes place.

When UNESCO (United Nations Educational, Scientific and Cultural Organization) started mapping desertification continent by continent, it drew up a series of four categories of severity which were then incorporated onto the maps. They were as follows:

- **Slight:** little or no degradation of the soil; little plant cover lost
- **Moderate:** (i) 26–50 per cent of the plant community consists of climax species
  (ii) 25–72 per cent of original topsoil lost
  (iii) soil salinity has reduced crop yields by 10–50 per cent

- **Severe:**   (i)   10–25 per cent of the plant community is of climax species
              (ii) erosion has removed all or most of the topsoil
              (iii) soil salinity has reduced crop yields by more than 50 per cent
- **Very severe:** (i)   Less than 10 per cent of the plant community is of climax species
              (ii) erosion has left the land with many sand dunes or deep gullies
              (iii) salt crusts have developed over irrigated soils which are barely permeable.

Areas in this last category are generally small and restricted and therefore not much in evidence on the continental maps. They represent land which has to be abandoned because it has been so badly degraded as to be worthless for agriculture. Desertification can be reversed in these areas but may not be financially viable in the short term in an LEDC without some outside help in the form of money or expertise. The other categories of desertification are much easier to reverse.

As well as desertification, the term **desiccation** is frequently used in the context of arid environments which are getting drier. It can in fact be regarded as a more extreme problem than desertification as it is more difficult to reverse. Goudie describes desiccation in the following way:

'It is the belief that portions of the Earth's surface are drying up leading to the spread of deserts, the depletion of groundwater reserves, the dwindling of rivers and the decline of settlements. By some it has been regarded as the result of progressive climatic deterioration in post-glacial times, while others have seen it as a result of the human mismanagement of the environment.'

Once again, the reasons can be said to be either natural or human induced, and it is easy to provide examples of both. There is plenty of evidence, such as that mentioned below, that the Sahara had a much wetter climate during the last Ice Age, and during the last 12 000 years it has become progressively drier; this is a process of desiccation. The catastrophe which has affected parts of Uzbekistan and Kazakstan bordering the shrinking Aral Sea can be regarded as an example of desiccation which has been induced by humans, and is dealt with later in this chapter.

3 Desertification as part of climatic change

Climate change is a natural process which has been taking place continuously throughout the history of the Earth. Over hundreds of millions of years (geological time), there have been huge variations in the climates of all parts of the world, with temperature changes of the magnitude of tens of degrees and rainfall changes of the magnitude of thousands of millimetres. The testimony of these changes lie in the rocks, and is particularly well illustrated in the stratigraphy of the UK. During the Devonian period (400–350 million years ago) for example, much of southern UK was experiencing desert conditions, the evidence for which is seen in the Old Red Sandstone rocks of Devon and South Wales; the rock salt deposits of Cheshire and elsewhere in the Midlands were also laid down during these arid times.

Over shorter periods of time, hundreds of thousands of years, the Earth has experienced moderately large climatic variations which have taken us in and out of the Ice Ages. The types of evidence which support these medium term changes in climate are geomorphological (evidence of past ice-cover), archaeological (human artefacts and animal bones), and botanical (pollen grain analysis).

During the last two or three millennia, and particularly in the last few centuries, historical records have enabled a much more detailed picture of climate change to be observed. Over this shorter timescale, average temperature variations of a few degrees between decades are not uncommon. Written records in many countries have provided evidence of what crops grew where and how good the harvests were.

The current debate on global warming and the worldwide impact of the extra greenhouse gas emissions of the last 50 years, should not obscure the simple fact that climatic fluctuations are a natural and normal part of the Earth's functioning, whether it be on a long term, medium term or short term. The possible reasons for these frequent, continuous and often erratic changes are many, and various physicists and geologists have their own theories. These theories may be put into three broad categories:

- those connected with changes in the Sun's activity and consequent fluctuations in the amount of incoming solar radiation
- those due to the planetary behaviour of the Earth, such as changes in the tilt of its axis and eccentric 'wobbles' in its orbit
- those connected to plate tectonics, such as 'global cooling' which follows periods of intense volcanic activity.

4 Evidence for climate change in arid environments

Deserts which have been hyper-arid for a long time, such as the Atacama and Namib, show few signs of climatic change. Rills and gullies carved by running water into the arid Atacama hillsides are fossil features in much the same way as those recently detected on the surface of Mars. The landscape of northern Chile has remained almost unchanged for so long that in places it is a 'dead' landscape, lacking all signs of geomorphic activity.

This is not true of the desert areas within parts of North Africa, the

Middle East, North America and central Asia, which were much more geomorphologically active in the recent past, and often continue to have high rates of change even today. Climatic change has had a considerable impact on these landscapes within the last few thousand years. Around 12 000–10 500 years ago, towards the end of the last Ice Age, many of these regions reached a climatic optimum with a much higher rainfall, and therefore experienced much more rapid rates of water action than they do today. This wetter period also stimulated the origins of sedentary agriculture in places such as Mesopotamia within the following two millennia. In addition to these longer term changes, there is plenty of evidence of short term fluctuations in rainfall. Evidence of climatic change takes a wide variety of forms and includes:

- **Geological evidence.** This type of evidence is found embedded in desert surfaces. Layers of gypsum, rock salt and other evaporites interbedded within water-deposited sediments such as clays are indicators of past drier periods. Cascades of tufa deposits which resemble petrified waterfalls may provide an indication of past wetter periods. Fossils of past plant and animal species also provide evidence of change within the rocks.
- **Geomorphological evidence.** The processes which mould the landscape will change as climates change. Sand dunes are generally mobile transportational features within deserts, but in areas of higher rainfall beyond the desert limits, formerly mobile dunes may be fixed. An increased rainfall brought about by climate change will allow plants to colonise dunes and thereby fix them.
- **Archaeological evidence.** Weapons, tools and bones found in archaeological sites provide evidence of past economies and food sources, which in turn may indicate wetter periods in the past. Similarly, the concentration of archaeological finds may be an indicator of past human population sizes; the larger the population, the wetter the climate. Prehistoric cave paintings also provide evidence of the types of wildlife that were hunted in the past, giving an indication of what the environment was like.
- **Historical evidence.** With the development of civilisations came the use of the written word by which important events (as well as everyday transactions) could be recorded. In ancient Egypt and Mesopotamia, records of harvests and food supplies were kept and these give insights into how variable rainfall and river floods could be from year to year. Longer term changes often led to the decline of one city state and the rise of another. Over the last two centuries meteorological records have become a new form of historical evidence; they are more complete in some desert realms than others.

When the evidence from these three groups of sources is put together, it becomes clear that climatic fluctuations in some, but not all deserts, can be very great over the long term, medium term and the short term. Over the medium term it is very difficult to interpret the role of humans in the degradation of the environment, as within arid areas their populations were generally limited and their economies well-adapted to the climatic conditions. In the short term, with rapid population growth, colonialism and the introduction of modern large-scale technologies, mistakes have been made which have upset the delicate balance of nature within arid areas.

5 The desertification debate

Whether or not parts of the world are suffering from desertification, whether or not such changes are reversible and where the responsibility for the process lies are all part of an ongoing debate within the scientific community. The first serious concern about the spreading of deserts in modern times was in the 1930s, when Stebbing published a paper entitled *The Encroaching Sahara: the threat to the West African Colonies*. Stebbing recognised the links between forest and savanna woodland clearance, soil degradation and the creation of barren land. He later regretted the idea of the 'encroaching Sahara' as it was taken up by other authors and became popularised to give the impression that there was a vast area of sand on the move which would engulf the lands to the south of the Sahara.

By the 1950s the spread of land degradation in the world had become a major concern of the newly formed United Nations;

CASE STUDY: EVIDENCE FOR CLIMATE CHANGE IN AFRICA

The continent of Africa has experienced many dramatic climatic changes over the last 1 million years, and evidence of this change is to be found both north and south of the Equator. Within present day deserts there is evidence of much wetter phases in the past, when conditions were much more similar to the savanna grasslands; similarly, within the present day savannas there is evidence of a much drier past.

Figure 6.1 shows many of the pieces of evidence of climatic change in Africa which are relevant to the arid and semi-arid areas of the continent, past and present.

Sand dunes

The map shows the distribution of mobile and fixed dunes within the continent. To the south of the 100 mm isohyet, stretching across the Sahel zone, most of the dunes are fixed today but were active in the past. In the Holocene period follow-

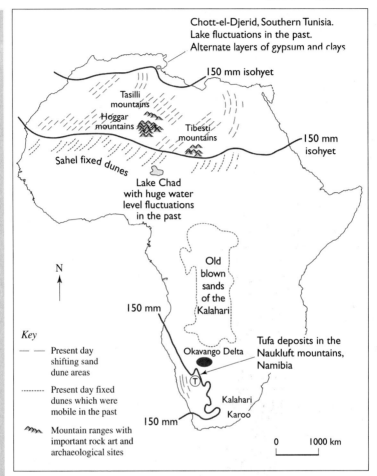

Figure 6.1 Evidence for climate change in Africa

ing the last Ice Age there appear to have been two major dry periods when dunes were active in the Sahel; the first of these would have involved a shift southwards of the main wind and rain belts by around 450 km, the second a shift of 200 km. The phase in between would have been much wetter, and like today the dunes would have been stable and subject to rain action carving out gullies and rills.

Blown sand deposits are extensive. Much of the dry savanna of central and southern Africa is underlain by thick Kalahari sands which extended northwards as a result of both a drier climate and unusually strong winds during the late Pleistocene period.

Lake level fluctuations

In wetter periods in the past, freshwater lakes were extensive in areas which today are occupied by much smaller chotts or totally dry depressions. Lake Chad provides one of the most striking examples of this. Its present surface is 282 m above sea level, but in the past it was as much as 400 m above sea level; this would have given it a surface area five times its present size. Ridges in the landscape provide evidence of past shorelines, and pollen grain analysis shows that Lake Chad was at its maximum size between 29 000 and 21 000 years ago and again between 14 500 and 7000 years ago.

Evaporite materials

Closely associated with lake levels are the evaporite materials found in and around areas which are or have been salt lakes in the past. At various places within the great intermontane basin of the Chott el Djerid in Tunisia, clays and mudstones are interbedded with gypsum and other evaporites. The sedimentary rocks mark the pluvial phases when freshwater lakes occupied the area or when it was an inlet of the Mediterranean. The evaporites mark the much drier periods such as the present when potential evapotranspiration rates are much higher than rainfall.

Tufa deposits

Deposits left behind by limestone solution may also provide clues to wetter climates in the past. The Namib appears to have been hyper-arid for many millions of years, but has occasionally in the recent past experienced short periods of higher rainfall, such as that which occurred 25 000 years ago. In the Naukluft mountains there is considerable evidence of one such period when the climate was wetter. These mountains, which form a high plateau, are made mainly of dolomite limestone and have far-reaching cavern systems within them. In some of the *kloofs* (the local name for narrow gorges) which incise the plateau there are deposits of **tufa** which resemble petrified waterfalls up to 30 m thick. These would have formed as a result of calcium carbonate deposits being laid down by solution-charged rivers which would have flowed when the water table was much higher than it is at present, indicating a much wetter climate.

Cave and rock paintings

In many arid parts of Africa rock paintings provide evidence of both human occupancy and the types of animals that were hunted during the recent past. In mountain desert areas rock art representations show that hippopotamuses, rhinoceroses, elephants, giraffes and other savanna species were common earlier in the Holocene period. Human artefacts such as the stone weapons used to hunt these animals are also found far from present day water sources. Some of the most striking evidence is found in the Tassili mountains in Algeria which experienced a wetter, grassland phase as recently as between 6000 and 2500 BC. Evidence shows that from 6000 to 4000 BC there were hunter-gatherers inhabiting the region and cave paintings show animals, and hunters with bows and arrows and exaggerated eyes as they watch their quarry. By around 4000 BC the hunters are replaced by dread-locked herders surrounded by long horned cattle and goats; wild animals such as ostriches, gazelles and giraffes are also present. After 2000 BC the depictions of savanna animals fade into history.

Historical evidence

A lot of information about climatic fluctuations in southern Africa has been pieced together from the writings of European explorers and settlers. Observations made of water levels in rivers, lakes and wells, the location of villages in relation to water supplies, combined with stories told by the old as to how the environment had changed within their lifetimes, are all important pieces of evidence. In the regions of the Karoo and Kalahari, and in particular the Okavango Delta system, there appears to have been a period which gradually became drier from the 1790s to the early nineteenth century, then during the 1820s and 1830s there were periods of more severe drought. The 1850s and 1860s appear to have been much wetter, but were followed by a decline into more frequent periods of drought towards the end of the nineteenth century. What this evidence illustrates well is that the experiences of the Sahel in the mid-twentieth century were almost certainly just part of the normal pattern of rainfall fluctuations which is found in the drier regions of Africa.

Meteorological and other scientific records

Accurate records covering a sufficiently large network of meteorological stations to provide enough data for meaningful patterns to emerge, are a comparatively recent phenomenon. In the last few decades it has been possible to draw up accurate patterns of climatic fluctuations within both the Sahara and the deserts of southern Africa. These have been particularly useful in the analysis of the Sahel droughts during the second half of the twentieth century, even though the network of weather stations was not particularly dense and there were some gaps in the data. The use of satellite technology and remote sensing have added greatly to the scientific data available for interpretation, especially when used in conjunction with instrumental readings on the ground.

UNESCO launched the first serious study of desertification in its 'Major Project on Scientific Research on Arid Lands' between 1951 and 1962. By the early 1970s there was a sufficient body of research for scientists to understand the basic relationships between variations in rainfall, poor land management and land degradation. This coincided with the 1969–74 period of drought in the Sahel which created an environmental and food crisis in six Sahelian countries: Mauritania, Senegal, Mali, Burkina Faso (then Upper Volta), Niger and Chad. It also coincided with a period of great debate concerning the unprecedented population growth in LEDCs, sparked off by such books as *The Population Bomb* written by the US biologist Paul Erlich. The Sahelian nations at that time had, and indeed still have, some of the highest birth rates in the world. This debate was given a continued airing during and after the next period of drought and famine in the Sahel, between 1979 and 1986.

In 1986 H E Dregne gave a rather angled definition of desertification in his book 'Desertification in Arid Lands':

'Desertification is the impoverishment of terrestrial ecosystems under the impact of man.

It is the process of the deterioration in these ecosystems that can be measured by reduced productivity of desirable plants, undesirable alterations in the biomass and increased hazards for human occupancy.'

In this definition the role of potential climate change had disappeared, the cause of desertification was seen purely as human activity. Works such as Dregne's supported the arguments of the neo-Maltusians such as Erlich, putting the blame for the crisis in the Sahel, and in other LEDCs which suffered from drought and desertification upon population growth. This enraged many officials from LEDCs. They saw that MEDCs, and in particular the USA, were criticising LEDCs for their high population growth rates and blaming them for disasters which were not just anthropogenic but also to do with natural changes. Moreover, MEDCs could be blamed for the uneven distribution of wealth in the world and not contributing enough aid to alleviate the problems encountered in LEDCs. This population versus the sharing of wealth debate between MEDCs and LEDCs continued and dominated the major UN Conferences in the 1990s, including the Rio Earth Summit of 1992 and the Cairo Population

Summit of 1995. The debate continued in 2001 with the refusal of the George W Bush administration to sign up to the Kyoto Protocol on the emission of greenhouse gases, on the grounds that LEDCs were given special concessions, which the USA did not want them to have.

Much more co-operation and understanding is needed between nations to sort out global issues such as desertification and land degradation.

6 The organisations dealing with desertification

One of the problems associated with desertification is the wide range of bodies dealing with it at different levels in different parts of the world. At the international level, various agencies of the UN are responsible for different aspects of the phenomenon. As mentioned above, the United Nations Convention to Combat Desertification (UNCCD) was set up in 1996, but long before this, two major UN agencies, UNESCO and FAO (Food and Agriculture Organization) were heavily involved in the study of desertification as well as coming up with some remedies for it. UNESCO is more engaged with the defining, mapping and monitoring of desertification, whereas FAO is more concerned with the practical investigations in the field and the remedial action which needs to be taken to halt the spread of the desertification process. Funding for various large-scale projects may come either from IFAD (International Fund for Agricultural Development) or directly from the World Bank, two other UN agencies.

An example of an institution which successfully operates at the regional level is the OSS (Obsérvatoire du Sahara et du Sahel), which was set up in Paris in 1992. This involves 21 countries, mainly French speaking, which are partially or entirely within the Sahara or its southern Sahelian extension. Its wide ranging activities include:

- to co-ordinate the national efforts of the member nations
- to establish which indicators provide evidence of the causes and effects of desertification
- to develop the information systems which monitor the various signs of desertification
- to achieve a better understanding of the mechanisms which lead to desertification, such as population shifts, the use of natural resources and environmental management
- to identify the methods and techniques that can improve natural resources and environmental management.

Each of the 21 nations has come up with a master plan for dealing with its own desertification problems and this involves international organisations, regional bodies, national ministries and NGOs as well as local 'grass roots' input (see the following case study).

CASE STUDY: DESERTIFICATION AND HUMAN RESPONSE IN TUNISIA

Tunisia, like its neighbours in the Maghreb region of North Africa is facing a considerable threat from land degradation, which could potentially in the long term lead to more permanent forms of desertification. The country covers 16.4 million hectares, of which 5.4 million are used for arable farming and a further 4 million are either forest or pasture. Only the northern 6 per cent of the country, categorised as humid and sub-humid, is not really at risk from land degradation. The Tunisian government has classified the remainder of the country as follows:

- (already) desert   21.5 per cent
- badly threatened by desertification   17.2 per cent
- moderately threatened   23.2 per cent
- slightly threatened   32.1 per cent

The threats have been summarised by the National Committee to Combat Desertification (CNCD) as being both physical and human. The physical threats identified are connected with climatic unpredictability and drought years alternating with wetter years, with wind and water erosion, and salination. The human factors identified are those which exacerbate the situation and are connected with population growth and greater demands being made upon the land. These include the decline of soil fertility, the lowering of the water table, the silting up of dams and the increased flood risk caused by new buildings and infrastructure.

Along the lines of those suggested by the UNCCD, Tunisia has created a National Action Plan involving the help of international, national and local organisations. At the international level are the UN institutions such as the FAO and UNCCD, at the national level there are the ministries such as those for Agriculture, for Economic Development and for the Environment, as well as research institutes, the ANPE (National Agency for the Protection of the Environment) and the CNDD (the National Commission for Sustainable Development). At the local level each region of Tunisia has its own development office, for which desertification and environmental protection are major concerns.

The work of the various Tunisian bodies in the fight against desertification has thus far been impressive. Action is taking place at various levels. Most important in the short term are the technical operations which involve the experts such as agronomists and hydraulic engineers working in the field. These operations include:

- Soil and water conservation, selecting the most appropriate farming techniques, terracing, the building of earth dams and general slope protection.
- Larger scale infrastructural projects such as dam construction across wadis to conserve water, floodwater distribution walls and stone or concrete erosion protection walls such as those made of gabions.
- A great variety of methods of sand dune stabilisation, including the provision of palm frond fences, the planting of hardy, fast growing trees such as eucalyptus and tamarisk both to stabilise dunes and to create wind breaks. Artificial ridges of dunes are also sometimes created to act as wind breaks.
- Reafforestation on certain areas of wasteland such as sebkhas and farmland which has become too saline.
- Chemical treatment of soils which are becoming too saline, deep ploughing and the adding of organic mater to counteract the alkalinity of degraded soils.
- The piecemeal addition of new fields to existing oasis areas thereby developing the area under plant cover, resulting in additional moisture and humus content in the soil.
- The recycling of drainage water from agricultural and domestic uses where possible, and the treatment of waste water at the domestic level for use in irrigation.

The second tier of projects to counteract desertification in the Tunisian National Action Plan are those known as horizontal projects, which involve many people including experts and farmers alike and provide an education and skills framework for action. These projects include:

A new oasis at Kebili in S. Tunisia, a part of the country's policy to halt desertification

- An inventory of soil resources and their suitability, through scientific soil mapping.
- An inventory of the plants of the region and their suitability for different climates and land regeneration at different levels.
- The development of a network of land degradation maps through remote sensing and the setting up of a national observatory of land degradation.
- The development of a guide to rational land management.
- The development of a drought readiness programme, which will involve extra water saving measures in times of drought, the conservation of animal fodder for emergency use and the use of agricultural forecasting techniques.
- Educational and public information schemes aimed particularly at the young.
- Changing and rationalising energy consumption in rural areas, particularly cutting down on the use of brushwood and increasing the use of gas ovens.
- The use of geothermal energy wherever it is available.

In order to see many of these measures through to fruition the government has implemented numerous support measures which are aimed at getting the public involved. These include training and public information schemes, support for research projects, the improvement of rural infrastructure and most important of all, the setting up of local branches of agricultural bodies co-ordinated by CNCD.

The Tunisian National Action Plan is well-devised and very comprehensive in the way it is setting out to deal with the problem. It is in many ways a model for other countries to follow, but Tunisia is lucky in being wealthier and more able to implement the plans than many African nations.

CASE STUDY: THE SHRINKING OF THE ARAL SEA

The setting

The shrinking of the Aral Sea represents one of the greatest acts of human folly of the twentieth century. At the centre of an internal drainage basin within the Turkestan Desert, the Aral Sea is located partly in Kazakhstan and partly in Uzbekistan; until 1991, however, before these countries gained their independence, they were both part of the old USSR. With an area of 68 000 km^2 and a volume of 1090 km^3, the Aral Sea was the fourth largest lake in the world until Soviet hydraulic engineers embarked on their large-scale water diversion schemes in the 1960s. The lake is only fed by two perennial rivers, the Amu Darya (Oxus River) and the Syr Darya, both of which rise to the east in the mountains of central Asia and discharge into the Aral Sea as a series of deltaic distributaries, which reflects the very flat, low-lying nature of the topography of the area. These deltas contained very rich and highly biodiverse marshes and wetlands covering some 550 000 hectares. The combined lake and delta fisheries produced local annual catches in the region of 40 000 tonnes.

Over-irrigation

This part of central Asia has practised irrigation for the last 2500 years, but it has only been in the last four decades that over-irrigation has led to the desiccation of vast tracts of land close to the Aral Sea. The dramatic changes in the hydrological inputs into the sea have resulted from both natural and anthropogenic events. There were periods of seriously depleted rainfall in the early to mid 1970s; again in the mid-1980s there were some drought years which lowered the flows in the Amu Darya and the Syr Darya by around 30 per cent. Much more disturbing than the natural change, however, was the huge amount of water extracted for irrigation. The lake has been used for millennia for irrigation, but it was only in the latter part of the twentieth century that extraction rates reached crisis proportions. In 1900, 3 million hectares were under irrigation and by 1950 this had been

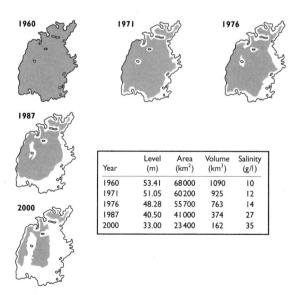

Year	Level (m)	Area (km²)	Volume (km³)	Salinity (g/l)
1960	53.41	68000	1090	10
1971	51.05	60200	925	12
1976	48.28	55700	763	14
1987	40.50	41000	374	27
2000	33.00	23400	162	35

Figure 6.2   The shrinking of the Aral Sea

extended to 5 million, yet the extraction of water had no real negative impact on the lake levels. It is since the 1960s that the human impact upon the Aral Sea has been greatest. The rapid development of irrigation in the central Asian republics of the old USSR was particularly directed at cotton production – and cotton is a thirsty plant which is also demanding on soil nutrients. The ecological problems of over-extraction began to appear in the 1970s. By 1980, the irrigated land had been extended to 6.5 million hectares, and for this purpose 132 million km³ of water was being extracted each year from the two feeder rivers – almost triple the figure for 1960. Oblivious to the damage that was being done, the Soviet authorities merely increased the rates of extraction throughout much of the 1980s, until in 1987 the irrigated area had been extended to 7.6 million hectares. Above all it had been the construction of the Karakum Canal, which stretches westwards for 1300 km taking irrigation waters out of the Amu Darya, that had the biggest impact on the desiccation of the Aral Basin. At the same time as local water supplies dwindled, the population grew from 14 million in 1960 to 27 million in 1980. Figure 6.2 shows the way in which the size of the lake has diminished as a result of excessive extraction of water for irrigation.

The problems created by over-irrigation

Eight major problems associated with over-irrigation and over-extraction of water can be identified in the Aral Sea region.

- **Climate change.** With the reduction of the size of the Aral Sea there have been significant changes in the climate of central Asia since the 1980s. The climate has become more continental, with hotter but drier summers and longer, colder winters with the growing season reduced to around 170 days per annum. Dust storms have become a greater problem, occurring on an average of 90 days per year. Although winters are colder, snowfall is less frequent and therefore the spring snowmelt input into the regional hydrological system has diminished.
- **The upsetting of the region's water balance.** Closely connected with local climate change is the upsetting of the regional water balance. The shrinking of the lake's surface has created a negative feedback mechanism. The decline in the surface water area has led to a decline in evapotranspiration, which in turn has led to a drop in precipitation levels and therefore surface run-off and groundwater flows which feed back into the Aral Sea. Through time this mechanism becomes a downward spiral of decline, with the amount of water caught up in the local hydrological cycle constantly diminishing.
- **Groundwater depletion.** The drop in the level of lake water has led to a significant drop in the water table within the Aral Basin. By

the 1990s, the 15 m drop in the lake level had resulted in a depression of groundwater levels by 7–12 m close to the lake shores and by lower amounts further inland. One major effect of the lowering of the water table has been the drying up of wells and the abandonment of once fertile farmland. Groundwater depression also leads to a greater potential for contamination and pollution of well-water.

- **The salination of the land.** The exposure of the former lake bed and the decline in groundwater levels both have a knock-on effect on soil quality and contribute to soil salinity. The exposure of the lake bed produces large salty crusts from which material can be blown onto productive farmland. The lowering of the water table leads to higher evaporation rates causing salts to be transferred upwards through the soils, leading to the formation of solonchaks. In the upper reaches of the basins of the Syr Darya and the Amu Darya only 10 per cent of the soils are rated as being either saline or very saline, whereas in the lower basins 95 per cent of soils are now in these categories.
- **Salt and dust storms.** About 27 000 km² of the former lake bed has become exposed. Much of this land is covered in concentrations of salts such as calcium sulphate, sodium chloride, sodium sulphate and magnesium chloride. Clouds of this toxic cocktail are blown over a wide area during windy periods; some deposits have been found as far away as Georgia in the Caucasus, over 1000 km from the Aral Sea. These salt and dust clouds which create storms up to ten times per year, are hazardous to human health and can poison the soil in the more fertile areas of the Aral Basin; 60 per cent of these storms cross the rich Amu Darya delta region.
- **The loss of agricultural productivity.** The decline in surface and underground water supplies has led to low irrigation efficiency. This has in turn led to sharp falls in agricultural productivity. Salination of land has led to the abandonment of as much as 40 per cent of the formerly irrigated area. The overuse of pesticides and artificial fertilisers has further polluted areas of arable land, rendering them useless for crop growth. With an estimated 81 per cent retreat of pasture lands through desiccation, livestock farming has gone into great decline.
- **Health hazards.** Desiccation has brought a host of health problems to the region. Urban and rural water supplies have become saline and polluted. In Karakalpakstan the water is polluted with a high concentration of metals such as strontium, zinc and manganese. In the last two decades there has been a 3000 per cent increase in chronic bronchitis and diseases of the liver and kidneys, especially cancer and a staggering 6000 per cent increase in arthritic diseases, as a result of pollution and desiccation. As a consequence of all this the infant mortality rates are some of the highest in the

world. This rapid increase in diseases has put a great strain on local hospitals and health services.

* **The depletion of biodiversity.** As the Aral Sea has shrunk, become shallower and more saline, the local biodiversity has diminished. 20 of the 24 lake fish species have disappeared and the commercial fishing ports of Aral'sk and Muynak are now found high and dry, a long way from the Aral shore. The deltaic wetland areas, where the Syr Darya and Amu Darya flow into the Aral Sea have decreased in size as a result of desiccation; they have consequently lost numerous species of plants, fish, birds and small mammals through the disruption of the food chain.

All of the problems listed above have contributed to the economic decline of the Aral Basin. In and around the lake itself, the major change has been in the retreat of the surface area of the Aral Sea and the consequent decline in the fishing industry. As the sea has shrunk, its salinity has increased to such a level (fourfold in two decades), that commercial fishing effectively ceased in 1982. Former fishing towns and villages now lie up to 70 km from the shoreline and the rusting hulks of fishing boats lie abandoned in the dunes of sand, salt and polluted dust.

In the deltaic regions of the two rivers many traditional activities have greatly declined. Commercial hunting and trapping of mammals such as the muskrat have largely disappeared; areas once used for livestock pasture are no longer viable because of the decline in natural hayfields.

In the once great productive commercial farmlands fed from the diverted waters of the Syr Darya and Amu Darya the economic decline is equally dramatic. The fields of the ex-Soviet central Asian agricultural 'miracle' are suffering from a desiccation process triggered by the grand schemes of Soviet hydraulic engineers. At least 40 per cent of the fields which once produced commercial crops such as cotton and tobacco have gone out of production. Furthermore the retreat of the land which is viable for arable farming has had a great impact on local food supplies.

The future and possible remedies

The problems of the Aral Sea are extreme; yet the desiccation of the Aral Sea region may be reversible in the long term. As early as 1982 the Soviet government had realised the importance of having a master plan, which would lay down strict extraction limits from the Syr Darya and Amu Darya. It was not until the break-up of the USSR in the early 1990s however, that matters were dealt with seriously. The five newly independent republics of central Asia set up a joint water commission to plan and regulate water resources in the Aral Basin.

Various international organisations have become involved in dealing with the problems, including the Food and Agriculture Organization of the UN and the World Bank. The International Fund for the Aral has resulted from multilateral and bilateral aid and study programmes. Among the projects under study are:

* a scheme to transfer freshwater from the Caspian Sea
* a much greater emphasis on the re-using of drainage water and treated waste water for irrigation
* the replacement of old irrigation schemes with more efficient ones
* the introduction of new crops, especially salt tolerant ones.

The current levels of irrigation are regarded as unsustainable, and given the shortage of local funding, attention is being placed on demand management rather than supply management. Any increases in the amount of land to be irrigated will be made very slowly over a long period of time. Water withdrawn within the basin has been stabilised at $110-120\,km^3$ per annum.

It has been estimated that $73\,km^3$ of water needs to be put back into the Aral Sea over a 20 year period for it to regain its 1960 levels. This would be very difficult to achieve under present circumstances, particularly as the regional climate is now drier than it was before.

Within Uzbekistan there has been considerable progress in bringing water back to the Amu Darya delta, some of the small lakes have been replenished, wildlife is becoming re-established and the fish stocks and catches are steadily growing. The nature of irrigation is being carefully examined and water saving techniques are being introduced where possible, although funding is always a constraint. Modern micro-irrigation methods will eventually replace the larger scale and more wasteful schemes of the past.

Only two republics share the Aral shoreline, but the other three former Soviet central Asian republics (Kirgizstan, Turkmenistan and Tajikistan) have headwaters of the Amu Darya and Syr Darya within their territories. Long term treaties signed by the Interstate Council for the Aral Problem countries envisage water extraction budgets for each country with different amounts allowed for wetter years and drought years. Afghanistan, which has 12 per cent of the Aral Basin within its territory, will also eventually be brought into the treaties.

Summary Diagram

Defining 'desertification'

Related phenomena

| Drought | Land degradation | Desertification | Desiccation |

Degreee of seriousness →

United Nations classification

| Slight | Moderate | Severe | Very severe |

Evidence of climate change

| Geological | Geomorphological | Archaeological | Historical |

Evidence in Africa

| Sand dunes | Lake levels | Evaporites | Tufa | Rock art | Historical | Scientific records |

CASE STUDIES

| Positive action in Tunisia | | The shrinking of the Aral Sea |

Questions

1. **a** Explain why the term 'desertification' is difficult to define.
   b With reference to specific examples and case studies examine the contention that desertification is irreversible.
2. **a** What are the main processes, physical and human which contribute to land degradation?
   b Outline the different ways in which land degradation can be counteracted or even reversed.
3. **a** Explain the differences between the terms drought, desertification and desiccation.
   b Outline the main ways in which human action can lead to land degradation in arid regions.
   c Using the statistics in Figure 6.2, assess the changes in the rate at which the Aral Sea has been retreating over the 40 year period from 1960 to 2000.
4. **a** Explain how desertification may continue to get worse in an arid area as a kind of vicious circle.
   b I low is this vicious circle changing the local environment in the Aral Sea region?
   c What remedies may be taken to slow down such a vicious circle?

7 The economic potential of arid and semi-arid regions

'The societies which are found in drylands in the twentieth century are the result of evolution over time. With some groups these changes have been very slow over hundreds of years and have resulted in a society which is in equilibrium with its environment. In other cases, changes have been so rapid in modern times that entirely new economies have been produced that are dependent not on local agricultural produce, but on fossil fuel subsidies.'

Beaumont (1989) *Drylands: Environmental Management and Development*

1 The nature of economic development in arid and semi-arid regions

Some countries which have deserts within their territory have a much greater percentage of arid land than others, and this largely determines what potential problems they face as they undergo economic development. Figure 7.1 categorises countries according to their percentages of arid land.

There has been a long history of colonisation, development of agriculture and evolution of cities in the arid regions of the world. Indeed, many of the great civilisations of the past were based on irrigated cultivation fed from the great rivers which passed through the arid realms of the Middle East and North Africa; the Mesopotamian civilisations, ancient Egypt and the cities of the Indus Valley all owed their existence to the human capacity for hydraulic

Countries that are 100% arid: Bahrain, Djibouti, Egypt, Kuwait, Mauritania, Oman, Qatar, Saudi Arabia, UAE, Western Sahara, Yemen.	Countries with 75–99% of their land classified as arid or semi-arid: Afghanistan, Algeria, Australia, Botswana, Burkina Faso, Cape Verde, Chad, Eritrea, Iran, Iraq, Israel/Palestine, Jordan, Kenya, Libya, Mali, Morocco, Namibia, Niger, Pakistan, Senegal, Sudan, Syria, Tunisia.	Countries with 50–74% of their land classified as arid or semi-arid: Angola, Bolivia, Chile, China, India, Mexico, Tanzania, Togo, USA.

Figure 7.1 Percentages of arid land within selected countries

engineering. In the Americas, from around the year AD 1000, in central Mexico and the South West USA, dense populations were supported in some arid areas by the use of irrigation schemes. The ability of arid regions to continue to develop, or at least sustain a growing population, has for centuries been based on their use of irrigation to maintain food supplies.

Although water supply is the single most important factor which determines the level of economic development in arid and semi-arid lands, abundant supplies of minerals have made certain desert states into some of the wealthiest nations in the world. Oil wealth has dramatically transformed the economies of the UAE, Kuwait, Saudi Arabia, Bahrain and Qatar in the last few decades. Libya, Iraq and Iran could also have been major economic powers within the Middle East and North Africa, had it not been for their political problems.

In other arid regions of the world such as the interior of Australia and northern Chile, it is the occurrence of other forms of mineral wealth which have enabled towns to be established in otherwise inhospitable areas. In such areas, however, a lot of capital has to be invested in developing water supplies and infrastructure to support the necessary labour forces to work the mines and transport the minerals to areas where they can be processed and exported. Such enterprises have to be profitable to be justified. The sharp fall in the world price of a mineral or the discovery of a new technology which might lead to it being replaced by something else, can cause the abandonment of mines and their settlements, which explains why so many 'ghost towns' are to be found scattered around desert areas.

Larger countries which have considerable lands outside of the desert realm are also fortunate in having a great deal of wealth generated outside of their arid lands by agriculture, trade and industry, and some of this wealth can be invested in the development of their deserts, which are often their 'last frontier'. This can certainly be witnessed in the USA, where the South West is the biggest growth area in terms of both wealth and population, and to a lesser degree in Australia's Outback and China's western provinces.

Desert climates may be hostile to human settlement, yet they have economic potential which leads to developments at many different levels; these will now be considered separately.

2 Aspects of arid region development

a) Nomadism, pastoralism and non-irrigated farming systems

Nomadism is the most flexible way of coping with an unpredictable rainfall and general shortage of water supply. Over thousands of years, nomadic ways of life evolved as a response to aridity.

Hunting and gathering economies were the first to be practised in arid lands as indeed they were elsewhere in the world. Only a few vestiges of these simple economies remain, and these have been altered by contact with the modern world. The various Aboriginal groups of the Australian Outback and the San of the Kalahari are examples of nomadic peoples who were traditionally engaged in hunting and gathering in arid lands, living in close harmony with their natural environment and making temporary migrations in response to any seasonal changes in food availability.

The majority of contemporary nomads in the arid realm are pastoralists. Where irrigation is not possible, nomadic pastoralism is the main human response to dryland farming. Scientists put the 300 mm or even the 400 mm isohyet as a minimum rainfall level below which pastoralism rather than cultivation should be practised. Nomadic pastoralism in drylands can be divided into three main categories: **pure nomadism**; **semi-nomadism** and **transhumance**.

Pure nomadism involves the use of temporary encampments and the total mobility of the herders who move with their animals from pasture to pasture and from water-hole to water-hole. The products of the camels, sheep, goats, cattle or other beasts provide the staple diet for the herders. Semi-nomadism involves seasonal movements with the flocks and herds, but during harvest time the pastoralists set up their tents or other forms of shelter near villages, and take advantage of the availability of supplies of wheat, sorghum and other staple crops to feed themselves and their animals. Transhumance is a form of 'vertical nomadism' practised in mountainous areas and is associated with mixed farming. Village women often tend to the crops while the men are engaged in the seasonal movements of the animals, taking them up into the high pastures in summer and bringing them down to lower pastures in the winter. In North Africa, the Atlas mountains are an area where transhumance is still widely practised.

In recent decades there has been great pressure in the Middle East, central Asia and North Africa to settle nomads down, or at least to restrict their movements. As populations grow and droughts become more frequent in many of the world's arid lands, there is increasing competition for land and it is the pastoralists who always seem to lose out. Although pastoralism is part of the deep-rooted culture of the Middle East and North Africa, it is generally looked down upon as an obstacle to development.

There are many reasons why nomads may settle down, some are voluntary and economic, others are political and forced. **Sedentarisation** as the process of settling down is known, can be for purely economic reasons when nomads embrace a more attractive or profitable lifestyle by moving to a village or a town. For example, the majority of the Toubou nomads of the Tibesti mountains in the Sahara have settled down on the Libyan oilfields. Their main source of income is now from working on the oil wells, but they still keep some livestock as a source of meat and milk. Similar patterns of sedentarisation can be seen throughout the Arabian Peninsula.

Forced sedentarisation has generally been carried out by governments in the name of development. In the former USSR republics of central Asia sedentarisation of large numbers of nomads was forced upon them by Stalin's regime of the 1920s and 1930s. In Niger in the 1980s there was a policy of sedentarisation which was not wholly enforced because of the refusal of many nomads to settle down. Sedentarisation may also be caused by more catastrophic events such as drought or civil war. In Ethiopia and Somalia large numbers of nomads have been introduced to a sedentary life as a result of time spent in refugee camps.

b) Irrigation and agricultural development

Nothing has been more important for the economic development of desert areas than advances in the technology of irrigation. The cultivation of arable land evolved in the Near East around 7000–8000 years ago, during what was known as the **Neolithic Revolution**. One of the events which stimulated this revolution was the retreat of the ice sheets in the Northern Hemisphere, which led to the desiccation of vast tracts of Arabia and North Africa. Irrigation was developed and expanded as a response to climatic deterioration. By around 5000 years ago great riparian civilisations based upon irrigation and food surpluses had evolved along the Tigris and Euphrates in Mesopotamia, the Nile in Egypt, the Amu Darya and Syr Darya in the Aral Basin of central Asia, along the

The great Ziggurat at Ur. One of the earliest cities developed within Mesopotamia using irrigation water from the Euphrates

Indus in what is today Pakistan and along the Huang He and Yangtze Rivers in China. Many of these locations were important **food hearths** where major crops which were found in the wild became domesticated staples, e.g. wheat and barley in the upper reaches of Mesopotamia and rice in southern China. Meanwhile, in the Americas, the forerunners of the Incas had developed gully and rill irrigation for other staples such as the potato, also by around 5000 years ago.

Traditional irrigation methods may have been used continuously for millennia in some parts of the world, yet in others frequent technical changes have enabled the constant improvements of crop yields.

The sources of water available for irrigation within desert areas have already been discussed in Chapter 3. They include the use of perennial rivers both exogenous and indigenous to deserts, deep and shallow aquifers, dams across wadis, the use of fog water and desalination. To these sources, large-scale **river diversion** needs to be added. In the former USSR schemes were planned to divert major rivers such as the Ob, which flow into the Arctic Ocean, to bring their waters to the deserts and dry steppes of central Asia. Although these schemes were never realised, the desiccation of the Aral Sea basin, dealt with in Chapter 6, provides the sort of evidence of the scale of environmental disaster that can result from diverting or over-exploiting large rivers. Also, schemes which are currently being created in Hussein's Iraq and Gaddafi's Libya by the building of huge 'man-made rivers' may prove to be ecologically dubious, both because of the amount of water they extract from perennial rivers or fossil aquifers, and the way in which they can speed up rates of soil salination.

i) Water collection and storage

Throughout the arid realm many different traditional methods of water catchment and storage can be identified. These include:

- The digging of basin-like depressions in the ground where water from wadis and floods may collect. Such basins include the *boulis* of the Cape Verde Islands and the *nadis* of the Thar Desert. Their practicality is limited by low storage capacity and high evaporation rates.
- Underground cisterns were widely used throughout the Mediterranean in the ancient world and continue to be used throughout the Middle East and South Asia today, e.g. the *tanka* of the Thar Desert. Although they are more expensive to construct than surface catchments, there is minimal evaporation loss.
- Water catchment management is of great importance in many countries. In many parts of North Africa wadi water is exploited by the construction of small dams of earth or stone, which divert moisture directly onto crops, e.g. the *bourj* system in Tunisia.
- Terracing has long been a method of water collection, as well as soil conservation in arid and semi-arid areas, particularly in mountain regions. Huge terraces survive from Inca times in the semi-arid parts of the Andes.

ii) Traditional irrigation techniques

Traditionally, irrigation methods have varied according to which water source is being used. Some techniques involve the digging of wells, others channelling, and others involve tunnelling. In some cases the waters may flow naturally and regularly, others trap irregular floodwaters and in many cases lifting devices are necessary because of changing water levels. Some of the most common traditional methods are as follows:

- **The use of perennial rivers.** The great civilisations of the ancient world used perennial rivers such as the Nile and Euphrates in numerous ways. During flood periods **basin irrigation**, which involved the flooding of shallow artificial embanked depressions, was very important. During other times of the year **perennial irrigation** was carried out by using various types of lifting devices such as the *sekia* or animal drawn water wheel. The development of irrigation in Egypt is discussed in a case study on pages 133–6.
- **Water diverting dykes.** In areas which have wadis rather than perennial streams, earth or stone built dykes can be constructed to divert flood water onto a cropped area either directly or through a network of canals. In the Draa region of Morocco this method is used to feed intricate networks of earth built canals called *séguias*.
- **The spreading of floodwaters.** Although less sophisticated than the

building of dykes, in many parts of the Sahel, water is collected in temporary ponds often with small dams behind them. This method does depend on there being a rainy season.

• **'Bour' cultures.** The word 'bour' refers to a tunnel shaped hole which is dug down to tap an aquifer. Date palms in Algerian oases are irrigated in this way from 15 m deep holes known locally as *soufs*.

• **Irrigation from springs.** Natural springs are one of the most important water resources for oases. Water issuing out of the ground from aquifers which come from higher, wetter areas have been used for millennia to feed intensely cultivated gardens through networks of artificial channels. The Nefta Oasis in southern Tunisia (see case study on page 64) is a good example of this type of irrigation.

• **Irrigation from wells.** Where aquifers do not reach the surface, their water has to be extracted from wells. All sorts of animals and human powered lifting devices can be used to bring the water to the surface to feed artificial channels to distribute the water onto the crops. The motor pump is now widely replacing the older types of lifting devices.

• **Irrigation from underground galleries.** One of the most sophisticated of traditional techniques is the construction of galleries which take water from aquifers on higher ground down towards plains. At various points there are wells down to the galleries to enable local water extraction, and the final destination of the water is generally an area of intense usage such as an oasis village. The galleries have the advantage of much lower evaporation rates than in surface canals. Extensively used in the Middle East these galleries are known as *qanats* in Iran and *falaj* in Oman.

iii) Modern irrigation techniques

The spread of modern technology is inevitably leading to a decline in the traditional variety of irrigation techniques. Many modern irrigation methods are much more efficient in terms of energy input, labour and water use. On the other hand they may require more capital outlay and thereby exclude small farmers from using them. At the same time, through what is known as **alternative** or **intermediate technology**, simple techniques can be upgraded by using modern materials thereby increasing the efficiency of irrigation for even the poorest peasant farmer. This is how the experience of farmers and engineers working in desert areas in MEDCs can be passed on to benefit farmers in LEDCs. Two major types of irrigation projects have become associated with new technology in the twentieth century:

• the construction of large scale dams, often a part of multi-purpose schemes
• the introduction of less wasteful hi-tech methods of watering the land.

There has been a great deal of debate about the building of large dams to extend a country's area of productive land. Early in the twentieth century it was regarded as a panacea by many countries, a key to their economic development. What was not envisaged at the time were the ecological side effects of the building of these 'mega-dams'. Since the last few decades of the twentieth century there has been a lot of criticism of projects on the scale of the Aswan Dam in Egypt, the Narmada Dam in India and the Three Gorges scheme in China. Such schemes have undoubtedly brought great advantages to LEDCs in particular, but in the long term have created whole sequences of negative knock-on effects.

Mega-dams often result in very wasteful use of water because of losses through evaporation and seepage from both their reservoirs and their large irrigation canal networks. One of the aims of modern irrigation is to increase the efficiency of water use. The schemes which are conservative with irrigation water tend to be those which are first developed with the financial and technological resources of MEDCs in desert areas such as the South West USA and the Australian Outback, which are then subject to **technology transfer** and used in other parts of the world. Given the wealth of certain oil-rich Gulf states, however, an increasing amount of the research and development is being carried out in the Middle East.

Some of the main forms of modern irrigation technology are:

- **Sprinkler irrigation.** This involves the use of pipelines rather than canals, thereby greatly reducing water loss from evaporation. It also enables great control of water flow and flexibility of where and when water can be applied to crops.
- **Central pivot or carousel irrigation.** This is a form of sprinkler irrigation where the pipes are mounted on wheeled frames which rotate through 360 degrees, spreading out the flow of water over areas of several hectares. The great green circles in desert areas of the Middle East such as in Saudi Arabia which are visible from the air were created by this form of irrigation.
- **Rapid flood techniques.** Some crops may need an occasional submersion in water rather than the more continuous flow from sprinklers. Rapid flood techniques release controlled amounts of water and can be electronically operated.
- **Drip feed irrigation.** Even sprinklers can be regarded as wasteful in areas with acute water supply problems. Drip feed involves a much slower but more continuous releasing of water from pipes and is often electronically operated.
- **Hydroponics.** Soils are generally poor in desert areas and hydroponics provide an answer to both soil and water problems. High value crops are grown in greenhouses in containers full of water with added minerals and nutrients, rather than soil.

c) Mineral wealth and economic development

Mineral wealth can lead to human settlement in areas where it could not otherwise be justified or sustained. The more valuable the mineral, the greater the resources which are used to develop the mining operation. Throughout the Middle East there are examples of how the presence of oil and natural gas have led to the transformation of nations' economies beyond nomadic pastoralism and oasis irrigation. The smaller states of the Gulf are dealt with as a case study on page 146 and exemplify how oil wealth can transform countries. Minerals other than fossil fuels are also important in desert areas of the world. The Outback of Australia, the desert areas of Namibia and the Atacama region of Chile all have considerable mineral wealth. Australia's Outback was partially opened up because of the gold rushes of the nineteenth century; there are still around 100 gold mines today in the ancient Darling Plateau of Western Australia, although activity is on the decline. Opal mining is still a major activity in Australia's desert areas as 95 per cent of the world's deposits are to be found in South Australia and New South Wales. Australia's most significant contribution to world mineral markets today is a range of metal ores, especially nickel, centred around the old gold mining town of Kalgoorlie, Western Australia, titanium (of which it is the world's main producer) and manganese.

Namibia, like most of Australia, has ancient crystalline bedrock which is rich in minerals. It has the world's richest diamond fields and a considerable area of the south of the country is off-limits to locals and foreigners alike as it is the property of Consolidated Diamond Mines, one of the biggest employers in Namibia. Ghost towns such as Kolmanskop bear witness to earlier diamond mining operations. As with Australia, Namibia is well endowed with some of the rarer minerals such as lithium, beryllium, vanadium and tantalum as well as more common ones such as tin, zinc and lead. Mining towns such as Tsumeb, Karibib and Rosh Pinah exist in hostile desert and semi-desert environments only because of the mineral deposits in their localities.

d) Manufacturing industries

Heavy manufacturing is an activity not normally associated with arid areas, particularly where a remote region is involved. Oil refining in the Middle East is particularly concentrated along the coastal plains, where land is flat, cheap and of little other use; also the exporting of the refined oil to other countries is facilitated by a coastal location. Industrial linkage has led to the development of other coastal industrial complexes in countries such as the UAE and Kuwait, which include petro-chemicals, iron and steel and construction materials. Ports may act as processing points for minerals extracted in their desert hinterlands, if the minerals are not too bulky or if they can be transported easily and cheaply to the coast. The phosphate processing works at Sfax on the coast of southern Tunisia is linked by railway to the mines at Metlaoui, 300 km inland. Similarly in Western Australia the iron ore from Koolyanobbing in the desert interior is brought by rail to the coastal steelworks at Kwinana near Perth for processing. Light industries are more diverse in arid areas and fall into two main categories: those that are specifically for local markets, and would be bulky to import, and those for which the desert location has been specially chosen such as hi-tech industries.

e) Scientific research and the space and defence industries

The clear anticyclonic skies, the remoteness of many desert regions and their low population densities have all made arid zones important to scientific research and the defence industry. Some of the world's major astronomical observatories are located in deserts, taking advantage of the almost constantly cloudless skies. Mount Palamor in California houses the Hale Observatory and also in the Californian desert is the Mount Wilson Observatory where the astronomer Hubble worked. On the Plains of San Augustin in New Mexico is the Very Large Array, a group of 27 telescopes which monitor the existence of distant galaxies. On the southern fringe of the Atacama, at 2200 m in the Cerro Tololo is one of the most important observatories in the Southern Hemisphere. Jointly run by the University of Chile and several US universities, it has the second largest telescope in the world. In Australia two of the most important observatories are in the semi-desert areas of New South Wales, at Mount Stromlo and Siding Spring.

Rocket launching, space research and nuclear testing are all particularly associated with desert environments. Much of the UK's early rocket and nuclear testing took place in the Australian desert. Woomera in South Australia was the UK's main rocket research station in the 1950s and 1960s, whereas Maralinga, a few hundred kilometres further into the desert, was one of the main nuclear testing stations. The latter has led since the late 1990s, to a dispute between the British government and Australian veterans' associations, as some 10 000 Australian servicemen were involved in nuclear testing and experiments on the effects of radiation on humans, and many have since died of cancer. In the USA the desert areas of the South West have many important military and research bases. Early tests of atomic bombs were carried out both underground and on the surface in the desert states, the first of these having been at Alamogordo, Trinity Mountain in New Mexico. Today the most important centre for research into both nuclear weapons and nuclear energy is that at Los Alamos, also in New Mexico. Large areas of the desert are in the hands of the federal government and off limits to the civilian population.

The situation is more complex in central Asia. In the former USSR, the most important locations for rocket launching and nuclear research were in the desert and semi-desert areas of Kazakhstan, in places closed to the rest of the world and not even marked on maps. Since Kazakhstan became independent with the break-up of the USSR in 1991, the existence of such sites is no longer a secret. The Baikonur Cosmodrome in the south near the Aral Sea was where most important launches of the Soviet space effort were made and today Russia still uses the site. Further north in Kazakhstan are 18 000 km² of semi-arid steppes at Semipalatinsk used between 1949 and 1990 for hundreds of nuclear bomb tests, below ground and in the atmosphere; this remains one of the most radioactive places on Earth, and a cause of political friction between Kazakhstan and Russia.

f) Tourism in desert regions

Deserts are rarely the single most important tourist attractions in arid countries. Many desert nations have coastlines, which, given an assured sunny climate all year round, have been developed as resort areas. Sousse and Monastir in Tunisia, Eilat on the Red Sea coast of Israel and the luxury beach hotels in Dubai in the UAE are all examples of beach tourism in arid or semi-arid areas. The long-established tourist industry in Egypt and the fledgling one in Libya are based essentially on the ancient civilisations and archaeological sites of those countries; if Iraq eventually finds itself able to develop tourism, it would be based partially on the Mesopotamian civilisations of its past.

The South West USA is one area where tourism is based firmly on the desert, its landscapes and its adventure potential. The ideas of the 'last frontier' and 'wilderness regions' are strongly embedded in the psyche of the USA for historical reasons. The National Parks and other areas of outstanding natural beauty in the South West USA attract millions of visitors each year, some on organised trips, but the vast majority are the American public driving, camping, trekking and indulging in adventurous activities in such places as the Grand Canyon, Lake Mead, Zion National Park and Bryce Canyon. In Australia, the desert interior (the 'Outback') is also an area which is an important part of the national consciousness and steeped in folklore. Although not developed to the same degree as in the USA, desert tourism in Australia follows a similar pattern, with some of the most visited areas of the interior including Alice Springs, Uluru (Ayers Rock) and Lake Eyre with its rich wildlife. Namibia has well-developed desert tourism, with its great variety of landscape types, but also because of its abundance of animal life, especially around the Etosha Pan, which like Lake Eyre in Australia is an extensive salt lake–salt flats complex. Namibian tourism had its foundations in the well-developed infrastructure built by South Africa during the time it was administered by its neighbour.

In most LEDCs it is the lack of infrastructure that keeps significant numbers of tourists away from desert areas; also political problems may act as a deterrent. Trans-Saharan treks were once popular with back-packers and adventurous off-road vehicle drivers alike, but the political problems in Algeria, the Polisario war in Western Sahara, the occupation of northern Chad by Libya and the ongoing civil war in Sudan have made virtually all routes impossible. Morocco, Tunisia and Egypt are all successfully adding the desert to their more traditional tourist locations.

Two divergent strains of desert tourism appear to be emerging: those which exploit the environment to the full and are in danger of destroying it, and those which treat it with the respect of the conservationist. The 'dune bashing' and the 'wadi bashing' activities of the UAE exemplify one extreme, where convoys of off-road vehicles travel at great speed across the desert landscapes with apparently little regard for the environment and its wildlife. In contrast on the Sinai Peninsula in Egypt there is a current trend towards a form of 'ecotourism' whereby visitors trek with the Bedouin and their camels, camping in the desert and living very much as the Bedouin have for centuries, causing minimal damage to the natural environment.

g) 'Sunbelt' retirement areas

The features that attract tourists to arid environments may also attract the retirement generation. The dry air and sunny conditions of the warm desert environment are generally much healthier than the conditions found in humid temperate regions, especially during the winter months. Certain illnesses which affect the elderly such as rheumatism and arthritis, bronchitis and pneumonia are less likely to be suffered in arid environments. In MEDCs there is a high degree of mobility of people at retirement age; those who migrate tend to be wealthy and their money stimulates the economy of the areas into which they move. In the USA, not only are the traditional rural and coastal locations attractive to the older generation, but also, certain desert towns are increasingly becoming centres of retirement. Phoenix in Arizona, Las Vegas in Nevada and Palm Springs in California are all examples of large 'oasis' cities in the US desert which have large concentrations of retired people. As yet this pattern has not been repeated in Australia, this being due to the country's low density of population and the availability of coastal areas for retirement.

International retirement to arid and semi-arid regions is also common. Large numbers of northern Europeans choose to retire to the Canary Islands in order to enjoy almost constant warm, dry weather throughout the year. One of the more remarkable 'sunbelt' retirement schemes was proposed in Japan in the 1980s. With a large ageing population and very high cost of living in Japan, various welfare organisations were proposing to build retirement homes on the semi-arid coastline of Senegal, where local people could be employed as nursing staff at a fraction of what it would cost in Japan. These proposals were not carried out, however, because of the downturn in the Asian economy in the 1990s.

h) The film industry in arid regions

Deserts, with their almost constantly clear blue skies and bright sunshine are ideal locations for film-making. Given their distinctive landscapes and landforms, desert locations lend themselves to a whole range of movie types from Westerns and science fiction to Biblical epics and other historical films recounting events which actually took place in deserts. Ever since John Ford's *Stagecoach* was made in Monument Valley on the Arizona–Utah border in 1939, the characteristic mesa and butte landscape of the valley has featured in many other Westerns, as well as other types of films and advertisements. The deeply gullied landscapes of Zabriskie Point in Death Valley, California were made famous by Antonioni's 1969 film of the same name, and it has become a well-visited tourist spot ever since. Through Westerns, 'road movies' and other films shot in the south western states, the desert has become part of the personality and perceived image of the USA. Similarly, the use of the Outback in films such as Stephan Elliott's 1994 *The Adventures of Priscilla, Queen of the Desert* has underlined the importance of the arid interior of Australia as part of the nation's personality.

David Lean's classic *Lawrence of Arabia* made in 1962 was filmed in some of the locations where the events actually took place in Jordan and Egypt. Similarly, although based on a work of fiction, the 1990 Bertolucci film *The Sheltering Sky* was shot in locations in Morocco, where the novel was set. Of all the countries in North Africa and the Middle East, Tunisia has been one of the most successful in attracting film crews; this is partly a reflection of its political and economic stability, in contrast with its two neighbours, Algeria and Libya. Various episodes of the *Star Wars* films of George Lucas were made in Tunisia from 1976 onwards, and in 1996 Anthony Minghella's *The English Patient* was also partly shot in the Tunisian Sahara. Film making can boost local economies in LEDCs such as Tunisia, through the fees charged to the film companies for the locations, through the employment of local people as extras and through the charges for local services, such as catering. Although this is generally only a form of short-term income, the movies may stimulate tourism and therefore a long-term income.

Some of the Sahelian countries, especially Mali and Burkina Faso are important for their own indigenous film industries, even though directors have to work with very limited budgets. Many of the themes of these films relate either to traditional stories handed down through the oral tradition, or to the problems of poverty and migration which the countries face today. The fact that the biennial Pan-African Film Festival (FESPACO) is held in Ouagadougou, reflects the importance of the role of Burkina Faso in the film industry.

CASE STUDY: IRRIGATION AS THE KEY TO EGYPTIAN DEVELOPMENT

There is a well-known saying that 'Egypt is the gift of the Nile'. Without the waters of the longest river in the world, Egypt would never have developed its ancient civilisation 5000 years ago, and the modern state of Egypt would not be able to support its current population of over 67 000 000. The Nile has two main sources, the White Nile which rises in the Equatorial regions near Lake Victoria, and the Blue Nile which rises in the Ethiopian Highlands (Figure 7.2). The former has a constant flow of water throughout the year, but the latter is much more seasonal, with its highest discharge in June–September, coinciding with the rainy season in Ethiopia. Within Egypt there are no perennial tributaries flowing into the Nile; it is essentially an exogenous river which brings water and life to an otherwise hyper-arid environment.

In ancient Egypt it was above all the floodwaters of the Nile, laden with rich alluvium eroded from the mountains, that enabled intensive agriculture to take place along a narrow strip of land on either side of the Nile's channel. It was the surplus food from this fertile land which enabled society to evolve and the great Pharaonic civilisation to thrive for 3000 years.

Until the great changes brought about by the building of the Aswan High Dam in the second half of the twentieth century, the traditional pattern of irrigation in Egypt took two main forms: basin and perennial irrigation. Basin irrigation relied on floodwaters, and earth banks were constructed to trap the floodwaters and silt into small field-basins in which new crops would be planted. The whole of the Nile Valley within Egypt benefited from this form of irrigation, but nowhere more than the Nile Delta where the river's distributaries spilled over to supply Egypt's largest agricultural area with water and silt.

Perennial irrigation, which relied on the river's normal water flow throughout the year, existed wherever it was possible to feed the canals which bounded the farmers' plots with water directly from the Nile. Numerous different lifting devices such as the human-operated pivoted bucket or *shaduf,* and the animal-operated waterwheel or *sakia* were the main methods of taking water from the river, especially important in the dry season. By the mid twentieth century, the diesel pump became widely used as a lifting device for those who could afford it.

Attempts to control the flood waters in a limited way by the building of barrages such as the Low Dam at Aswan, were made by the late nineteenth century; these enabled permanent towns and villages to be built closer to the river's channel. The biggest

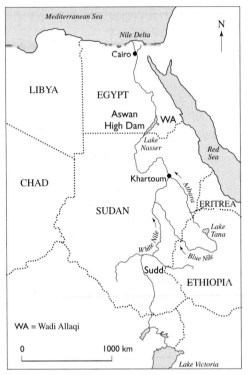

Figure 7.2 The Nile Basin

change came in the early 1970s with the completion of the Aswan High Dam, which radically altered the regime of the Nile throughout Egypt. Effectively the Nile no longer floods, the discharge of the river fluctuates very little from month to month, and the whole country must now rely on perennial irrigation. Behind the High Dam is the reservoir, Lake Nasser, which holds on average, 108 billion m^3 of water. Given the expansion of perennial irrigation, Egypt increased its agricultural output threefold within 15 years of the dam being built and the practice of double-cropping became commonplace.

The building of the dam has also proved to have had negative side-effects which are both ecological and economic. Figure 7.3 shows how some of these problems have arisen. The lack of new silt which used to come down with the annual flood is the root of many of these problems, because artificial chemical fertilisers now have to be used in large quantities.

Whereas a lot of the farmland continues to be worked in much the same way as in the past, there are many changes taking place in agriculture and irrigation techniques in contemporary Egypt. Even at the very local level, the complicated and rather unwieldy wooden *sakia* has almost completely disappeared and has been replaced by a simple mass-produced metal version, this being a good example of well-applied appropriate technology. There is a great divide appearing between traditional and modern farming in Egypt, and this has parallels in other North African countries such as Tunisia and Morocco. At the traditional level is the *fellahin* or peasant farmer, still producing subsistence crops, albeit with an increasing amount of produce grown for the local market. At the modern level is the large-scale production of cash crops such as vegetables, cotton, sugar and rice. This is particularly the case in Lower Egypt in the Delta where the vast urban markets and industrial plants lie close by in Cairo, Alexandria and the other large cities. In regions of the Delta such as El Sahir, central pivot sprinkler irrigation is widely used in the production of high yielding horticultural produce. These rotating sprinklers create circles of cropland 60 hectares in area, reminiscent of agricultural developments in richer Middle Eastern countries such as Saudi Arabia. The use of water by this method is at least 50 per cent more efficient than the traditional techniques, although it involves much more capital

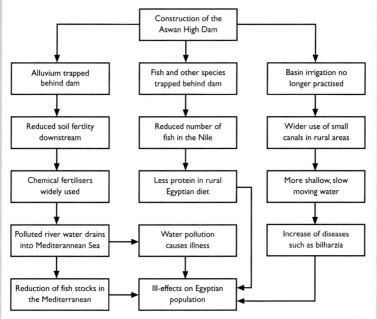

Figure 7.3 Ecological problems caused by the Aswan High Dam

outlay; this means it is generally practised only by the larger landowners. The combination of Egypt's all year round sunny climate, the introduction of hi-tech irrigation, and the expansion of large business organisations within the agricultural sector, have meant that Egypt is now a major producer of fruit and vegetables, such as tangerines, peppers, aubergines and new potatoes for European markets.

The future development of irrigation in Egypt depends on finding and tapping new sources of water, the more efficient use of existing water resources and any further changes that may be made upstream on the Nile. The wadis which flow into Lake Nasser are one potential new irrigation source, as water seeps out of the lake into them. A pilot scheme is at work on Wadi Allaqi (see Figure 7.2) where a new farming community has been established on the shores of Lake Nasser. More efficient use of present water supplies can be made by increasing high-tech irrigation on large-scale farms and introducing new forms of appropriate technology for the smallholders.

The biggest problem for Egypt in the future will be the decisions made upstream by the Ethiopian and Sudanese governments. Neither of these countries has yet made much use of the Blue or White Nile for large-scale irrigation. Both countries have plans for dams across the rivers which could reduce the flow of water downstream to Aswan. Ethiopia provides the greater threat as it has the sources of 11 tributaries that feed into the Nile and provide 80 per cent of the water which eventually reaches Egypt. Now that the country has emerged into a period of peace and stability following famine and a long civil war, it has proposed over 40 new irrigation schemes, many of which would involve dams across the Blue Nile or its tributaries.

In order to try to prevent future conflicts on this issue, the Nile Basin Initiative (NBI), which is backed by World Bank funding, was embarked upon in 1999. Following a conference in Dar es Salaam in Tanzania, ten countries including Egypt, Sudan and Ethiopia agreed to co-operate on the use and development of the waters of the Nile Basin. A Council of Ministers representing the ten countries now meets on a regular basis in order to discuss future schemes. The success of the NBI could influence the political stability of this region, which has many problems and conflicts, as well as having a rapidly growing population which will make increasing demands on its limited water resources.

CASE STUDY: THE ECONOMIC POTENTIAL OF SOUTHERN TUNISIA

Tunisia has a North–South divide, which is based mainly on the availability of water. The northern half of the country, with its Mediterranean climate and annual rainfall ranging from 600 to 1500 mm, has much greater agricultural potential in high yield crops such as olives, grapes, almonds and citrus fruits than the more arid south. The north is the economic core of Tunisia; the higher rainfall has historically given it a higher density of population as well as a greater degree of urbanisation. The large cities of the north such as Tunis, the capital, Bizerte and Sousse are ports which are major centres of trade and industry. Their position, closer to Mediterranean trading nations of the EU, such as Italy, France and Spain, has also enabled Tunisia's north to expand its economy more rapidly than the south. Recent expansion of the Tunisian economy has been based on inward investment with the setting up of Export Processing Zones (EPZ) by transnational corporations, which take advantage of the country's cheap labour force. The vast majority of this EPZ activity has, however, also been in the north and has widened the economic gap between north and south even further.

One of the major consequences of this economic disparity has been the heavy internal migration from the more traditional agrarian societies of the south towards the industrialised cities of the north which have much higher standards of living. This internal migration pattern is uneven, however, with many more moving away from the Tataouine and Médenine regions of the south east than from the Tozeur and Kebili regions of the south west (see Figure 7.4). Given the high rates of unemployment throughout Tunisia, (15 per cent national average, but higher in the south), there is also considerable emigration to European countries.

Government action through the *Commissariat Général du Développement Régional* (General Commission for Regional Development) is aimed at evening out some of the disparities between north and south by making huge investments in all the main economic sectors.

Despite the water shortages, the Tunisian south is well endowed with minerals. Tunisia's largest mineral deposits are the phosphate fields around Metlaoui; the main oilfields are located in the far south near the Libyan border and off-shore in the Gulf of Gabes; salt lakes such as the Chott-el-Djerid are another source of mineral wealth. Five of Tunisia's top ten industrial companies are based on the mineral wealth of the south; four of them petrol companies, and the fifth (Compagnie des Phosphates de Gafsa) is engaged in mining. These may bring

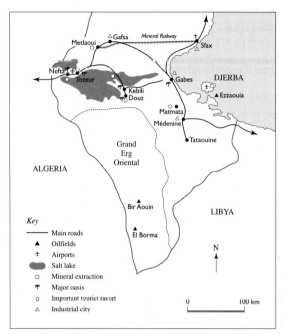

Figure 7.4 Southern Tunisia (Economic)

investments to the south but are not necessarily large-scale employers. Heavy industry is concentrated in the largest city of the south, Gabes, which has a population of 650 000, and includes the processing of phosphates mined at Metlaoui, cement works for the construction industry and various types of engineering works. Where EPZs have located in the south, they have mainly been around Gabes; the low wages they offer have encouraged female employment and have done little to reduce male unemployment.

Most of the industrial employment in the south is still in small-scale enterprises such as food canning and packaging plants, brickworks and workshop industries producing goods such as textiles and household utensils, mainly for local consumption. Tourism is providing an increasingly large market for certain workshop industries such as rug-making and ceramics.

The majority of people in the south still work on the land, and this is strictly limited by what water resources are available. Nomadic herding, as elsewhere in North Africa is on the decline, due both to economic changes and government settling down policies. Semi-nomadism, with people attached seasonally to villages at harvest time is now much more widely practised. There are advantages to this as sheep and goats produce a much higher quality of meat when fed on the better pastures found near villages. As most agriculture is now arable, water supply is crucial; there are few naturally occurring surface sources in the south and the water table has been lowered rapidly in the last few decades. In order to keep up with agricultural demands and a growing population, a huge number of new schemes are being embarked upon, these include:

- New dams, especially around the larger towns. The high evaporation rates are an important factor in their design.
- More solid and durable small-scale wadi catchment schemes in rural areas to harvest the rainfall more efficiently and transfer it directly onto the crops before much evaporation loss.
- New, deeper boreholes in the oases and other rural areas. This has often led to the bringing up of groundwater at 60–80 °C, which then has to be cooled in elaborate fountain and tower structures before it can be applied to crops.

The agricultural sector in the south is expanding as a result of the use of these new water sources and through specialisation. This takes two main forms: the concentration on successful traditional crops and the introduction of new high value crops. In the oases of Tozeur, Nefta, Kebili and Douz, dates are still the major cash crop. Much of the newly irrigated land surrounding the old oases is used for the production of horticultural produce, with crops such as peppers, chillies and aubergines grown for export to European markets. New labour intensive farms such as these are to some extent stemming the outward flow of migrants.

Tourism is also adding a boost to the economy of southern Tunisia. The country is seen as the safest in North Africa, and with its short travel time from Europe has developed its coastline as a destination for package holidays. With the exception of the island of Djerba, which receives some 200 000 visitors a year, tourism is much more concentrated in the north, in such resorts as Hammamet, Sousse and Monastir. The biggest growth area for tourism is, however, the desert. Oasis towns such as Tozeur and Douz have set up 'Zones Touristiques' with numerous luxury hotels and other tourist infrastructure. These zones were set up with large grants from the central government and the hotels are given a five year tax-free period once they open. Not only do they provide employment for unskilled workers, but the new college for hotel management and catering, opened at Tozeur in the late 1990s, is giving the southerners a chance to train for high level jobs within the industry. By exploiting the tourist potential of the desert, Tunisia is giving another dimension to what European visitors can experience within the country, and in addition to creating thousands of new jobs, it has stimulated traditional crafts and enabled many camel herders to find a new role for their animals and earn a much higher income than in the past. In Douz, for example, 90 per cent of the traditional camel herders now work in tourism. Some argue that tourism is bad for the south because, as most luxury hotels have swimming pools, it puts great pressure on valuable water supplies; in Tozeur this has been further exacerbated by the construction of a new luxury golf course. Another criticism is that tourism is causing the local culture to be replaced by a 'Disneyland meets Arabian Nights' version of it, designed to entertain the tourists. These problems are perhaps the price which has to be paid for bringing greater prosperity to these oasis towns.

As well as direct employment in the tourist industry there are many jobs indirectly connected with tourism, such as in gift shops, restaurants, bars, and taxi and bus drivers. As money is invested in the growing oasis towns and the local populations expand, further employment is generated in local service provision, in schools, hospitals and administration.

CASE STUDY: THE ECONOMY OF NORTHERN CHILE

Chile is an extremely elongated country which stretches for 4000 km from north to south along the eastern Pacific, yet it is barely 250 km in average width from east to west due to the way in which it is hemmed in by the natural barrier of the Andes mountains. Chile has three distinctive climatic regions: the sparsely populated Atacama Desert in the north, a central zone with a Mediterranean-type climate and a much denser population known as the Central Valley, and a cooler and wetter southern region of mountains and fjords, with limited population and economic activity.

The three most northerly of Chile's twelve administrative regions: Tarapaca, Antofagasta and Atacama (see Figure 7.5) are the only ones which lie entirely within the desert. Although the nation's economic core is in the Central Valley, where Santiago, the capital, and Valparaiso, the main port are both located, a great deal of Chile's original wealth was based on the riches of the desert. It was the mineral wealth of these regions which attracted the Spanish *conquistadors* to this part of South America in the mid sixteenth century.

Water shortages are severe within the Atacama as the region has the lowest rainfall totals on Earth, a negligible amount of

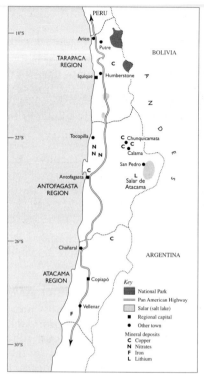

Figure 7.5 Northern Chile

surface run-off and large tracts of saline land where evapotranspiration rates are extremely high. This means that agricultural activity is very limited. In places in the foothills of the Andes agricultural production appears to have been greater in the past, as is evident from old abandoned Inca cultivation terraces. Many of the rural areas of the three northern administrative regions of Chile have experienced severe depopulation over the last century, with people migrating to the coastal towns and mining centres where the major economic activity has been concentrated.

In the absence of water, the two main economic resources of the Chilean Atacama are the cold Humboldt Current – one of the reasons for the region's hyper-aridity – and the mineral deposits in the desert. Many of the world's most productive fisheries are to be found in cold waters. The upwelling of the cold waters of the Humboldt Current along the coastline of northern Chile provides enough oxygen to support a huge population of plankton which provides food for a very large fish population. Tarapaca region, based on the ports of Iquique and Arica, has 65 per cent of Chile's industrial fisheries. Catches are processed on

Humberstone, a ghost town in the Atacama abandoned in the 1940s
when synthetic nitrates were developed

the spot for export in freezing plants and canneries, with crustaceans and anchovies as two of the most important products. The region is also a great exporter of processed fishmeal made from a wide variety of species and used both for animal feed and fertilisers. The Antofagasta region, based on the major port city of the same name, produces a further 10 per cent of the nation's fish, but to its south, the Atacama region is only of minor importance to fisheries as the Humboldt Current's influence on water temperatures is not as great there.

Northern Chile was once nicknamed 'The World's Chemistry Laboratory' because of the range and concentrations of minerals in the Atacama Desert. Two main types of mineral wealth are found in northern Chile, those associated with the high evaporation rates of the desert and those associated with the zones of mineralisation within the contorted Andes. A third source of minerals, now largely worked out, were the once abundant coastal deposits of *guano* dropped by the large seabird colonies on cliffs and offshore islands; these nitrate rich deposits were extracted for use as fertilisers. The Spanish came to Chile in search of gold and silver but were disappointed, but it was not until after Chile had gained its independence from Spain that the mineral wealth started to be exploited. In the nineteenth century the world's richest deposits of *caliche* or sodium nitrate were discovered; this mineral is a valuable resource in the manufacture of fertilisers and gunpowder. The origins of these deposits can be put down to past volcanic activity as well as the hyper-arid climate. Throughout the north, boomtowns grew up on the *caliche* and between the 1880s and 1920s it accounted for over two thirds of Chile's income. In the 1920s a synthetic version of sodium nitrate led to the decline in demand and many of the Atacama mining towns closed down. Only two are still in production today and eerie ghost towns such as Chacabuco and Humberstone lie abandoned in the desert.

Northern Chile has many other minerals which continue to keep its primary sector economically buoyant. Copper is currently the most important of the resources extracted from the rich zones of mineralisation within the Andes, and accounts for up to 40 per cent of the country's export earnings. At Chuquicamata the largest open cast copper mine in the world is located with a 430 m deep and 3 km long pit. The mining operations support a local population of about 150 000 people.

The populations of Chuquicamata and Calama have stimulated the development of agriculture in the small oases of the surrounding districts. The most important of these, San Pedro de Atacama, is also developing as a centre for tourism. A huge range of natural phenomena such as the Valley of the Moon, a rugged landscape carved out of rock salt, the Salar de Atacama salt lake, active volcanoes, and the El Tatio geyser fields are all within easy travelling distance from San Pedro. The town also has fine Spanish colonial architecture, Inca ruins as well as a museum with Inca gold and pre-Colombian mummies which are well-preserved because of the hyper-aridity of the Atacama. Tourism along the coast is limited to a few main resorts, the main ports such as Iquique and Antofagasta having luxury hotels along their best suburban beaches. Tourism is mainly national rather than international, although new luxury resort complexes developed by transnational corporations are beginning to appear between Antofagasta and central Chile.

Despite having only 8 per cent of Chile's population the 'Great North' has very good infrastructure because of its mineral wealth. Most of the old railways have been abandoned, but the road network based on the Pan-American Highway is as modern and well constructed as in most MEDCs. Antofagasta and Arica are expanding their ports with modern container facilities, alongside the more traditional bulk handling of mineral exports. The main cities of northern Chile are modern and bustling with high rise buildings and shopping malls, reflecting the wealth of the Atacama mineral resources which has been invested in them.

CASE STUDY: ECONOMIC DEVELOPMENTS IN SOUTHWEST USA

With the exception of some areas of the Middle East, the USA has the most economically developed tracts of desert in the world. Fifteen of the western states of the USA have vast areas of arid or semi-arid lands within them, but the core of the arid South West is formed by the five contiguous states of Arizona, Colorado, Nevada, New Mexico and Utah. The water supply for this region comes mainly from the Colorado River and its various tributaries and its water problems have already been dealt with in Chapter 3 in the case study on pages 39–41. Sustainability of the economic growth and lifestyle in South West USA are looming as key issues for the future.

Although the South West remains one of the least densely populated parts of the USA, it is the region which is experiencing the fastest population growth. The states of the so called 'Sun Belt' are attracting new inhabitants partly because of the wide range of jobs being created there in new hi-tech and service industries, and partly because of their healthy climate which is ideal for retirement. Phoenix, Arizona is one of the most rapidly growing cities in the USA. In 1950 there were just 100 000 inhabitants, but today there are over 2.5 million people living there. This phenomenal growth has put a strain on water resources, yet the city has an exceptionally high standard of living for an unfavourable natural environment. Given the prodigal use of water for both irrigation and domestic purposes (well over 50 per cent of households have swimming pools), each inhabitant effectively consumes 1000 litres of water per day, compared to just 200 litres a day in the average European city. Although a smaller city, Las Vegas in Nevada and many other cities in the South West are equally extravagant with their water supplies. The American writer Susan Arritt puts the problem thus: 'Sophisticated technologies that enable people to inhabit and exploit dry lands may be costly to the environment, as well as to the economy, and – like all human-made devices – subject to malfunction. Consider the impact of a long-term power failure or water system breakdown on a subtropical desert metropolis. Touch of a button relief from the heat has clouded 20th-century perceptions of the inherent harshness of arid environments. It is a mistake to assume that, in the long run the desert can be tamed'.

Long before European settlers arrived in the South West there were numerous tribal groups of North American Indians living in mud built 'pueblo' settlements based on irrigation. By the time Europeans arrived these villages had been largely abandoned and nomadic tribes such as the Apache and the Navajo were the main inhabitants of the South West. Although some 'pueblo' Indian villages survive in Arizona and New Mexico today, the biggest populations in the past appear to have been in Colorado, where at the Mesa Verde archaeological area alone there are 1800 settlement sites.

European settlers in the area were first interested in primary resources, and the south western states are still very important producers of minerals, as well as timber in the more mountainous areas. Copper is the major mineral resource, especially in Arizona, Utah and New Mexico, whereas Colorado has the biggest coal reserves in the USA. Silver and gold mines are scattered throughout the region and rarer minerals such as mercury and beryllium are abundant in Utah. The evaporite crusts of the Great Salt Lake in Utah have processing plants which extract a range of salts including potassium sulphate and sodium sulphate for export throughout the world.

The aridity of the area meant that the European settlers introduced cattle ranching as the first major farming activity and Colorado still remains one of the top cattle farming states today. The development of more intensive forms of agriculture in the region came with the harnessing of the waters of the Colorado Basin throughout the twentieth century. Now huge areas of crop specialisation, such as cotton, grapes, nuts and vegetables are found in the five states as a result of the gradual extension of the irrigation network. However, this has not been without problems such as salination of land, lowering of water tables and conflicts over water rights.

The manufacturing industry of the South West falls into two main categories: those which are more traditional and related to the region's natural resources, and those which are there for strategic reasons. In the first category comes the smelting of minerals such as copper, the processing of foodstuffs and the paper, printing and publishing industries which have grown as a result of the local timber industry. The remoteness of the region, its low density of population and generally clear skies – all products of aridity – have made the South West important for defence and associated industries. In Nevada 85 per cent of the land is under Federal ownership. The Nevada Test Site for testing nuclear missiles was set up in the 1950s and this led to a concentration of a whole series of defence related industries in the Las Vegas area including aerospace, biomedical environmental protection and electronics. Air bases in the state are major employers in the tertiary sector. New Mexico, where the first ever nuclear bomb test was carried out at Trinity Mountain in the White Sands Desert in 1945, has a similar clustering of defence industry activities with the Los Alamos National Laboratory for nuclear research and offshoot private firms producing armaments, electronic equipment and precision instruments. Arizona and Colorado also have concentrations of defence related industries and military installations. The South West of the USA is one of the most heavily radioactively polluted places in the world. Not only are the remnants of the nuclear tests buried 1500 m under the desert surface, but also the Yucca Mountain site in Nevada is likely to become the dumping ground for thousands of tonnes of high grade nuclear waste from all over the USA.

Despite this, tourism and retirement are two rapidly expanding sectors in the South West. In Nevada tourism employs 35 per cent of the workforce. As well as wilderness areas with recreational facilities like Lake Mead, the state has the unique phenomenon of Las Vegas and Reno, oasis cities with their economies based on gambling, luxury hotels and live entertainment. The other south western states have a huge range of different types of attractive desert scenery, some of them with National Park (NP) or National Monument (NM) status. These locations include the Grand Canyon NP in Arizona, the most visited natural site in the USA, Bryce Canyon NP and Zion NP in Utah, and the White Sands NM in New Mexico. Some of these sites are of cultural significance too, such as the Mesa Verde NP with its 'pueblo' Indian villages, in Colorado. Where the Rocky mountains are high enough in Colorado, Utah and New Mexico there are numerous winter sports resorts, so tourism in these states is well spread throughout the year.

Retirement is closely related to tourism as they are both attracted to similar types of location. The south western states have a rapidly growing retired population because of the dry climate, year round sunshine and potential for sports and other outside activities throughout the year. Not only are the bigger cities attracting retired people but there are specific settlements designed for 'senior citizens' such as Sun City near Phoenix and Green Valley near Tucson.

CASE STUDY: THE SMALLER OIL-RICH STATES OF THE ARABIAN GULF

The four small states located along the southern shores of the Arabian Gulf (see Figure 7.6) would probably not be economically viable as independent nations if it were not for their oil wealth. Kuwait, Bahrain, Qatar and the United Arab Emirates (UAE) (itself a union of seven smaller states), have all undergone a rapid transformation in the last few decades because of

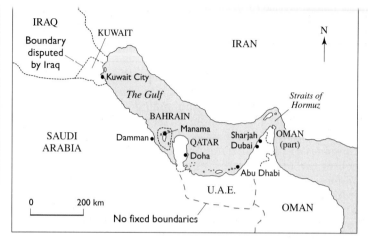

Figure 7.6 The smaller states of the Arabian Gulf

their large oil reserves and the world's heavy dependency on Middle Eastern oil. Before the discovery and exploitation of oil, all of these places were quiet backwaters on the Gulf, trading between Arabia and the Indian sub-continent; they were small harbour towns with fleets of *dhows* strung out along the coast whilst the interior was inhabited mainly by nomadic pastoralists. Before oil, the main source of income of most of these ports was pearl fishing.

All four are now active members of OPEC (Organisation of Petroleum Exporting Countries) and therefore are involved in the fixing and regulation of world oil prices to their own benefit. All four also have between 90 and 100 per cent of their territories technically classified as desert, which is potentially a problem for further economic development outside of the oil industry. All four are still heavily reliant on oil and natural gas as the mainstay of their economies: in Kuwait oil accounts for 93 per cent of the export income, in Qatar 81 per cent of the export income comes from oil and natural gas supplies, in Bahrain oil accounts for 60 per cent of national income, and only in the UAE where the figure is 40 per cent, do oil exports account for less than half the national income.

The Gulf states' oil reserves are not unlimited, for example in Qatar it is estimated that oil supplies will run out by the middle of the twenty-first century, and therefore all four of them have been taking action to diversify their economies, particularly in the last 20 years. Each state has approached diversification in a different way and has had a varying degree of success. Political issues, in particular the Gulf War of 1991, and religious attitudes towards westernisation have also had an impact on the way in which development has taken place. All four states have embarked on some degree of industrialisation, in some cases connected to the oil industry, e.g. petrochemicals, and in other cases using non-oil based local or imported raw materials. Given the large sums of money available from oil wealth, as well as the willingness of oil-consuming MEDCs to bolster their economies, there is no shortage of investment capital in the Gulf states. The biggest growth area in the economies of the four states, particularly in the last decade, has been in the tertiary sector, especially focusing on banking and commerce. Comparable to the European microstates such as Andorra and Liechtenstein, these smaller nations have found economic success by setting up duty free trading areas and 'offshore' banking. Tourism is another tertiary activity with great potential, although only the UAE has thus far developed its tourism on a large scale.

Kuwait

Kuwait with a population of 1.9 million and a per capita GNP of US$20 470, is the richest of the four states. It has an estimated 150 years of oil reserves, this being about 10 per cent of the world's total reserves. By 1990 it had developed a thriving oil-based economy. Although 95 per cent of the revenue from oil was in government hands, the people benefited from a model welfare state funded by this money. Everyone had free healthcare and heavily subsidised electricity, water and telephone services. Banking and the main heavy industries, such as petrochemicals and cement were also in state hands, and the Kuwaiti ruling family saw little need for foreign investment. In August 1990 the Iraqis invaded and occupied Kuwait until expelled in February 1991. Oil production ceased, 800 oil wells were set alight, much of its infrastructure was badly damaged and the economy was in ruins. An ecological disaster area was predicted in the Gulf as millions of barrels of oil were deliberately dumped by the Iraqis into the sea.

Both the waters of the Gulf and Kuwait's economy recovered much more quickly than had been feared. As a result of the Gulf War the Kuwaitis liberalised their economy more, allowing greater foreign investment and letting foreign banks locate within the country. Kuwait remains a highly controlled economy, however, with much of the financial power still in the hands of the ruling family. Corruption scandals involving ministers and members of the royal family, such as that of the unofficial stock market, rocked the Kuwaiti economy during the mid-1990s and to some extent held back the process of diversification.

The vast majority of Kuwait is desert and only 0.3 per cent of its land is under cultivation. The country was one of the pioneers of seawater desalination, which it started to develop in the 1950s. Today, intensive agriculture is highly concentrated in pockets of land, especially close to Kuwait City itself and based on desalination plants which provide over 75 per cent of Kuwait's water, at a rate of 400 million litres per year. Kuwait also pioneered the practice of hydroponics as early as the 1960s. Livestock, date production and citrus fruits are three areas where Kuwait produces food for export.

Qatar

Qatar is on a peninsula joined onto Saudi Arabia and has 12 per cent of the world's oil reserves and a per capita GNP of US$18 400. As with the neighbouring states, the oil wells were developed by foreign-owned petroleum companies, such as the Anglo-Iranian Oil Corporation and Royal Dutch Shell, but the production was then later taken over by a nationalised corporation (in this case the Qatar Petroleum Authority). Oil wealth has been invested in diversification of the economy in a number of ways. The fishing industry was revitalised and large heavy industries such as cement, iron and steel ship repairing and fertilisers have also been developed by oil investments. Advantage is taken of the cheap energy resources in the development of heavy industry. The creation of so many new manual jobs led to heavy immigration from poorer parts of the Islamic world, to such an extent that only 20 per cent of the 600 000 people resident in the country are Qataris.

Qatar has an impressive welfare system with free education and healthcare for all its people; the country also has no income tax. Although rather rigid in the past, Qatar's economy is being opened up to greater foreign investment, particularly in the banking and tourism sectors. The success of the UAE's economy has been a stimulus to this liberalisation process. Doha, the capital is located on the east coast, which has long sandy beaches and coral reefs and several new luxury hotels have recently been built there.

One of the most dramatic developments in Qatar has been the agricultural sector. Oil wealth has enabled the setting up of numerous desalination and waste water treatment plants which provide irrigation water for over 400 intensive farms; these take up only 0.7 per cent of the national land area. Qatar is self-sufficient in many fruits, vegetables and other foodstuffs and is an exporter of produce such as tomatoes to some of its neighbouring countries.

Bahrain

Bahrain is the Gulf state which is most aware of the limited nature of its fossil fuel reserves. This group of 33 small islands, with the main island (Bahrain Island) joined on to the Arabian Peninsula by a causeway, has a population of 600 000 and a per capita GNP of around US$12 000. Bahrain is likely to run out of oil towards the middle of the twenty-first century, but has more extensive supplies of natural gas. It has an economy which is more open to foreign investments than either Kuwait or Qatar and has also been quicker to diversify because of its dwindling oil resources; only 60 per cent of its income is now from fossil fuels. The nationalised oil industry not only processes its own petroleum but also feeds its vast refineries with imported crude oil from neighbouring Saudi Arabia. The main heavy industry developed with oil income is the ALBA (Aluminium Bahrain) works, the largest non-oil complex in the Gulf. Bahrain was one of the first places in the Gulf to allow foreign electronics firms to establish assembly plants. Smaller scale industries which supply the local markets include food and drink processing, plastics and furniture.

Three per cent of Bahrain's small area is taken up by agricultural land, irrigated from springs and desalination plants, although some of the best potential farmland is where the heavy industry is located. A wide range of fruits and vegetables are produced on an intensive and hi-tech basis, mainly for home consumption, but some for export. The onshore fisheries, particularly the shrimp farms are another important element in Bahrain's food production.

Bahrain's more liberal economy has enabled it to develop the tertiary sector to a high degree. In the 1970s and 1980s when the Lebanon was in turmoil, Bahrain took advantage of the situation and partially replaced Beirut as a major banking, insurance and commercial services centre for the Middle East. Tourism is not highly developed despite Bahrain's more open society. There is certainly great potential for tourism given the limited area of the country; Bahrain has sandy beaches, coral reefs, historical Islamic towns and archaeological sites. Manama, the capital, has a wide range of hotels as well as traditional bazaars and modern shopping complexes.

UAE

The UAE, with 10 per cent of the world's oil reserves and a per capita GNP of US$17 400, has the most open of all the Gulf state economies. The seven emirates, and in particular the three largest, Abu Dhabi, Dubai and Sharjah all have a rapidly expanding tertiary sector; Dubai has even been dubbed the 'Hong Kong of the Middle East' because of its commercial success. The seven emirates had been, along with Qatar and Bahrain, part of the Trucial States which had been administered by Britain prior to their independence in 1971. Qatar and Bahrain decided to go it alone. Whereas most of the hydrocarbon production and oil reserves lie within Abu Dhabi, where the political capital is located, most of the non-oil developments are concentrated in Dubai, which is effectively the financial and commercial capital. Other emirates are also undergoing greater specialisation of function; Sharjah is becoming increasingly important as a manufacturing centre, and a new free port area is being developed at Umm al-Qaiwain.

Of the 2.8 million people in the UAE only 25 per cent are of local origin, the remainder being immigrants, such as Egyptians, Pakistanis, Bangladeshis and Iranians who moved there in search of work, first of all in the oil industry, then in other sectors as the economy has become successfully diversified. Only 40 per cent of the national income is from its oil industry, a further 40 per cent from commercial activities and other services, with the remaining 20 per cent being accounted for by manufacturing and utilities. The most important heavy industries in the UAE are petrochemicals, aluminium smelting, fertilisers and construction materials; the last of these is particularly significant given the amount of infrastructural expansion the country is undergoing, with new roads, airports, container port facilities, hotels, shopping centres and office blocks.

Apart from the occasional range of hills such as the Djebel Hajar, the UAE is low-lying and the coastal areas are dominated by *sebkhas* or salt flats. The amount of arable land is limited to just 0.5 per cent of the country's area, but there are high density cultivated areas along the coast fed by desalination plants, in the oases such as Al Ain and in the irrigated valleys of the Djebel Hajar. As well as the traditional products of the oases, including dates and other fruits, hi-tech cultivation of vegetables using such techniques as hydroponics is now taking place, as well as very successful dairy farming.

As the UAE has a mixed state and private enterprise economy, it has allowed foreign banks and investors to operate there for several decades. Foreign investment has led to the setting up of the huge free-trade zone at the port of Djebel Ali, where 735 international firms have located, in conjunction with vast trade fair facilities and conference centres. One of the main aims of this expansion of the service sector is for the UAE to achieve a non-oil per capita GNP of US$20 000 by the year 2010.

Tourism is one of the UAE's fastest growing sectors. Given the more liberal nature of some of the emirates, particularly Dubai, international tourism has been expanding there in a way which is not possible in other parts of the Middle East. The UAE has numerous attractions, good infrastructure and is strategically

A modern commercial complex in the CBD of Dubai, UAE

positioned as an international transport hub between Europe and places further to the east. Oil money has been invested in the development of two world-class airlines, Gulf Air and Emirates Air which both have extensive networks. The UAE offers people a whole variety of different types of holidays from stop-overs of one or two nights, 'city breaks' of a few days to week or fortnight long packages which include beaches, coral reefs, sporting activities with world-class facilities, historic town centres and forts, duty free shopping and excursions into the desert. One of the more unique activities for which the emirates are famed is their various camel racing events. The luxury end of the tourist market is undergoing rapid expansion. In the late 1990s the hotel capacity on the articially created Jumeirah Beach in Dubai was greatly increased with the opening of the 'space age' 321 metre high Burj el-Arab Hotel. Currently a much more ambitious development is taking place. Two huge resort islands, shaped like palm trees, are being created on reclaimed land. Each will house up to 40 hotels and 1000 villas and as Sultan bin Sulayem, chairman of the developers states: 'The island environment will be similar to that of perhaps the Bahamas, Mauritius or the Maldives.

Summary Diagram

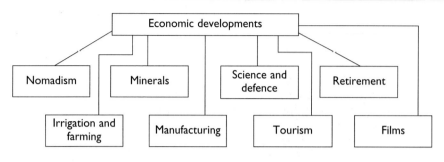

- Economic developments
  - Nomadism
  - Minerals
  - Science and defence
  - Retirement
  - Irrigation and farming
  - Manufacturing
  - Tourism
  - Films

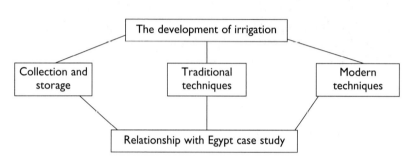

- The development of irrigation
  - Collection and storage
  - Traditional techniques
  - Modern techniques
  - Relationship with Egypt case study

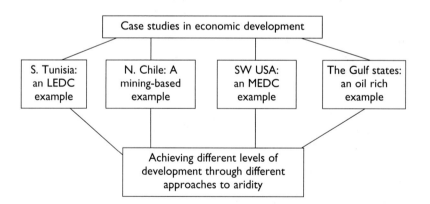

- Case studies in economic development
  - S. Tunisia: an LEDC example
  - N. Chile: A mining-based example
  - SW USA: an MEDC example
  - The Gulf states: an oil rich example
  - Achieving different levels of development through different approaches to aridity

Questions

1. **a** Outline some of the main obstacles to economic development in arid areas and some of the ways in which they may be overcome.

   b With reference to specific regions and countries, assess how successful people have been in overcoming the difficulties presented by arid environments.

2. **a** Explain the role of irrigation in the economic development of different desert areas.

   b How have irrigation techniques changed in recent decades and what impact has this had on economic development?

3. **a** Outline the main factors which might give an arid country economic potential?

   b With reference to specific examples, suggest what advantages MEDCs have over LEDCs in the development of their desert areas.

4. **a** Explain why some economic activities are better suited to arid lands than others.

   b Under what conditions can an arid area undergo economic development even if it has a severe water shortage?

   c Using the material in the case study on pages 146–52, outline why some oil rich states have been more successful than others in broadening their economies.

Bibliography

Adams, W., Goudie, A., and Orme, A., 1996, *The Physical Geography of Africa* (Oxford: OUP)

Beaumont, P., 1989, *Drylands: Environmental Management and Development* (London: Routledge)

Bradshaw, M., 1981, *Earth: The Living Planet* (London: Hodder and Stoughton)

Capon, B., 1995, *Plant Survival* (Portland, Oregon: The Timber Press)

Clark, J., and Bowen-Jones, H., 1981, *Changes and Development in the Middle East: Essays in Honour of WB Fisher* (London: Methuen)

Cook, R., Warren, A., and Goudie, A., 1993, *Desert Geomorphology* (London: UCL Press)

Errazuriz, A-M., 1992, *Manual de Geografia de Chile* (Santiago: Editorial Andrés Bello)

Gerrard, J., 2000, *Fundamentals of Soils* (London: Routledge)

Goudie, A., (ed.) 1997, *The Human Impact Reader* (Oxford: Blackwell)

Goudie, A., and Watson, A., 1980, *Desert Geomorphology* (London: Macmillan)

Goudie, A., 1984, *The Nature of the Environment* (Oxford: Blackwell)

Hill, M., 1999, *Advanced Geography Case Studies* (London: Hodder and Stoughton)

IGM, 1994, *Atlas Geografico de Chile para la Educacion* (Santiago: Instituto Geografico Militar)

Mainguet, M., 1999, *Aridity, Droughts and Human Development* (Berlin: Springer)

Mielke, H., 1989, *Patterns of Life* (Boston: Unwin Hyman)

Newson, M., 1992, *Land, Water and Development* (London: Routledge)

Olivier, W. and S., 1989, *Visitors' Guide to Namibia* (Cape Town: Southern Book Publishers)

Powell, J., 1895 (reprint 1961), *The Exploration of the Colorado River and its Canyons* (New York: Dover Publications)

Sale, C., 1985, *Australia: the Land and its Development* (Canberra: Australian Government Publishing Service)

Seely, M. 1992, *The Namib: A Shell Guide* (Windhoek: Shell Namibia Limited)

Thomas, D., (ed.) 1989, *Arid Zone Geomorphology* (London: Bellhaven Press)

Van der Merwe, J., (ed.) 1983, *National Atlas of South West Africa* (Cape Town: University of Stellenbosch)

Walton, K., 1969, *The Arid Zones* (London: Hutchinson University Library)

Periodicals:

The following periodicals frequently have articles on deserts or desert countries: *The Geographical Magazine; Geography; Geography Review; Understanding Global Issues; The Journal of Arid Environments* (Academic Press) – devotes all its articles to deserts

Websites:

There are numerous websites relevant to desert studies, some excellent, others less useful.

Five groups of types of sites are particularly useful:

- United Nations sites: www.fao.org www.worldbank.org www.unesco.org
- Specialist sites within the UN: www.unccd.int www.nilebasin.org
- Sites set up by individual countries: e.g. www.tunisiaonline.com www.ins.nat.tn both for Tunisia
- University Geography Department sites e.g. www.strathclyde.ac.uk/ Departments/Geography

Index